# HUMAN IN THE LEAD

*The Executive Playbook for Safe, People-First AI in Financial Services*

**SMIT JANI**

Human in the Lead: The Executive Playbook for Safe, People-First AI in Financial Services



ISBN: 979-8-234-14680-9 (paperback)

First Edition, 2026
Printed in the United States of America

# CONTENTS

# Foreword

*Every leadership team in financial services is having the same conversation right now.* It starts with questions that sound technical (which models, which vendors, which use cases), and it ends somewhere else entirely, with questions about trust, accountability, and people. Who is responsible when an algorithm makes a decision that affects a financial outcome for our business clients? How do we move quickly without breaking confidence that took decades to build? What happens to our people, and what do we owe them?

Those are not technology questions. They are leadership questions. And that is precisely why this book matters. I have spent my career in products and services firms, much of it leading teams through waves of change that were each, in their moment, declared revolutionary. Some of those revolutions delivered. Others quietly disappointed. The difference was almost never the technology itself. It was whether the institution adopted it in a way its people, employees and clients alike, could understand, trust, and use. Artificial Intelligence will follow the same rule, at far greater speed and with far higher stakes.

Sharing my experience of working with these firms and Fortune 100 clients, I can say that technology does look like the alpha in the equation, but it is human intelligence, governance, and insights that drive most of the innovation across each build. Whether we talk about the human in the loop, the human on the loop, or the human who helped engineer the loop, these interventions help deliver superior outcomes in every scenario.

What distinguishes *Human in the Lead* from the shelf of AI books written for executives is that it refuses the false choice between speed and safety. Smit Jani has spent fifteen years inside transformation programs at financial institutions across North America, and it shows. The frameworks here (the three models of human oversight, the governance foundations, the ninety-day roadmap) were not theorized at a distance. They carry the marks of real deployments: the pilot that stalled because the front line didn't trust it, the governance review that found oversight happening on paper but not in practice, the program that succeeded because someone asked the unglamorous questions early.

The book's central conviction, that AI should amplify human judgment rather than replace accountability for it, sounds obvious stated plainly. It is anything but obvious in practice. Taken seriously, it changes how you structure teams, what you measure, how you face your regulators, and how you talk to your own people about their futures. The chapters ahead turn that conviction into something executives rarely get from books on this subject: a plan you could begin executing on Monday. Pay particular attention to the chapters on human oversight and measuring people-centric returns. The first dismantles the comfortable fiction that putting a human in front of a screen constitutes meaningful control. The second answers the question every CFO eventually asks, "What is all this governance actually worth?" with a framework that treats trust as the asset it really is.

Artificial Intelligence will keep evolving faster than any of our planning cycles. The institutions that thrive will not be the ones that adopted it first, but the ones that adopted it in a way their clients and employees could believe in. That is the discipline this book teaches, and it is the right discipline for this moment.

Keep the human in the lead. The pages that follow show you how.

Sanjay Kukreja
CTO, eClerx
Mumbai, 2026

INTRODUCTION

# Why This Book, Why Now

The question has been circling her mind for weeks before she finally brings it to the table.

It's a Tuesday morning. The CEO of a mid-sized bank sits at the head of a glass-walled boardroom on the 32nd floor, looking around at her executive team. She's run this institution through a financial crisis, a pandemic, and two core banking conversions. She's not someone who loses sleep easily. But for the past three weeks, she has been lying awake at 2 a.m., turning over a question, she can't quite frame, one that feels less like a technology problem and more like an identity crisis for the organization she's spent twenty years building. She takes a breath, sets down her pen, and asks it:

> *"Are we falling behind on AI?"*

The silence lasts about three seconds. Then everyone talks at once.

The Chief Technology Officer starts listing platforms, vendors, and proofs of concept. The Chief Risk Officer raises a hand and asks who approved the model running in consumer lending. The Head of Wealth Management wants to know why the AI assistant his advisors were promised six months ago still isn't working. The Chief People Officer hesitates, glances down at her notes, and then says quietly that three high performers in operations just resigned, citing fears about "being replaced by robots." The CFO, arms crossed, asks a question nobody wants to hear: "What exactly have we gotten back from the twelve million we've spent so far?"

The CEO watches the crosstalk build. She notices that nobody is looking at anyone else; they're all staring at their own slides, their own spreadsheets, their own concerns. And in that moment, she realizes that the problem isn't the technology. It's that every person at this table is experiencing AI as a threat to something they care about (their team, their budget, their reputation, the people they manage), and nobody has given them a shared way to think about what comes next.

This scene, or something very like it, is playing out right now in executive suites across the financial services industry. And it reveals something important: the biggest challenges around AI in banking and wealth management aren't technical. They're human.

They're about trust. About leadership. About the people who build AI, the people who use it, the people it affects, and the people who are supposed to govern it. Get those things right, and the technology works brilliantly. Get them wrong, and you end up with expensive experiments, regulatory headaches, and an organization that's more confused than when you started.

That's what this book is about.

• • •

## Finance Runs on Trust: AI Can Strengthen or Break It

Let's start with something we all know but don't say often enough: financial services is fundamentally different from other industries.

When a streaming service's recommendation algorithm suggests a bad movie, you shrug and pick something else. When an e-commerce site shows you an irrelevant ad, you scroll past. These are inconveniences. They don't change your life.

Now imagine a woman named Anika, sitting at her kitchen table at 11 p.m., staring at a mortgage denial email she doesn't understand. She has a steady income, a solid credit history, and she's been saving for this house for four years. What she doesn't know is that an AI model declined her application based on patterns it learned from data that encoded decades of inequality, patterns that have nothing to do with Anika and everything to do with a system no one fully examined before deploying it.

Or picture a retired teacher named David whose bank account is suddenly frozen while he's visiting his grandchildren abroad, because a fraud detection system flagged a transaction as suspicious, and no human at the bank can explain why or override the decision quickly enough.

Or consider Maria, a small business owner who faithfully paid her insurance premiums for twelve years, only to have her hurricane damage claim auto-denied by an AI system that misclassified satellite imagery of her property. The appeal process is automated too, and weeks pass before a human adjuster finally reviews her case, weeks during which her shop sits waterlogged, and her employees have nowhere to work.

Or consider an algorithmic advisor that recommends a portfolio allocation to a client named James without accounting for the fact that he's three years from retirement and caring for an elderly parent with mounting medical bills.

These aren't inconveniences. These are life-altering moments. Anika, David, Maria, and James are not edge cases; they are the reason financial services exists. They are people who have entrusted an institution with something deeply personal: their financial security. Their savings. Their home. Their retirement.

This is why the stakes around AI in financial services are so much higher than in most other sectors. When we talk about AI in banking and wealth management, we're not just talking about efficiency gains and cost reduction, though those matter. We're talking about decisions that affect real people in ways they may feel for years. We're talking about a sector where trust is the product, as much as any loan or investment or insurance policy.

AI has extraordinary potential to strengthen that trust. It can help advisors know their clients better. It can catch fraud faster. It can make compliance more consistent and lending more accessible. But it can just as easily undermine trust, through opaque decisions, biased outcomes, impersonal experiences, or governance failures that erode the confidence customers and regulators have in your institution.

The difference between those two outcomes isn't primarily about the technology. It's about the choices we make about how to design, deploy, and govern that technology. It's about people.

• • •

## Why Most AI Programs Stall or Fail

Here's a pattern I've seen play out dozens of times across the industry. See if this sounds familiar.

A bank or wealth management firm decides to get serious about AI. Leadership is enthusiastic. Budget is allocated. A talented team is assembled. They build something impressive in the lab, a model that can predict customer churn, automate a compliance workflow, or generate personalized advice at scale. The proof-of-concept demo goes well. Executives nod approvingly.

And then... nothing happens.

Well, not nothing, exactly. What happens is a slow, grinding descent from excitement to frustration. The model doesn't integrate cleanly with the legacy core banking system. The compliance team raises concerns that nobody anticipated. Frontline staff don't trust the AI's recommendations and quietly work around them. Customers complain that the new chatbot is worse than the old phone menu. The project gets quietly shelved, or limps along as a "pilot" that somehow never graduates to production.

This isn't a hypothetical. It's the norm. The data tells the story clearly: as of early 2026, virtually every financial firm surveyed planned to put AI agents into production, yet barely one in ten had managed to do so at scale. Far fewer reported measurable financial benefits from their AI investments. The gap isn't ambition; it's execution.

And when you look at why these programs stall, a clear pattern emerges. The failures are almost never purely technical. The algorithms work. The data, while messy, is manageable. The cloud infrastructure is available. What breaks is everything around the technology:

- **Governance gaps:** Nobody clearly owns the AI's outcomes. Business, technology, risk, and compliance each assume someone else is responsible.
- **Change resistance:** Employees fear for their jobs or simply don't trust the AI's outputs, so they work around it rather than with it.
- **Customer friction:** AI-powered experiences are designed to reduce costs rather than serve customers, and the difference shows.
- **Regulatory blind spots:** Teams build first and worry about compliance later, only to discover that "later" means months of rework.
- **Measurement myopia:** Success is defined exclusively in terms of efficiency (cost savings, processing speed, headcount reduction) while trust, adoption, and fairness go unmeasured.

Every one of these failure modes is a people problem, not a technology problem. And yet, most AI strategies treat people as an afterthought, something to worry about after the model is built, the platform is chosen, and the architecture is designed. That's backward. And it's why so many of these efforts fail.

• • •

## What "People-Centric AI" Actually Means

So, what's the alternative? What does it look like to put people at the center of AI strategy in financial services?

Let me be clear about what I'm not saying. I'm not arguing against AI. I'm not suggesting we slow down innovation or bury every initiative in committee reviews. I'm not anti-technology. Quite the opposite; I believe AI will transform financial services more profoundly than any technology since the internet, and the institutions that don't embrace it will fall behind.

What I am saying is that the way we embrace it matters enormously. And the most successful approach (the one that leads to AI that actually scales, earns trust, and delivers sustainable value) starts with people.

People-centric AI means putting human judgment, accountability, and relationships at the center of how we design, deploy, and govern artificial intelligence. It's built on a few core convictions:

- **AI should augment people, not just automate tasks.** The goal isn't to remove humans from the equation. It's to make humans better at the things that matter most, building relationships, exercising judgment, showing empathy, and navigating complexity. In practice, that means an advisor who spends less time on data entry and more time listening to a client describe what retirement actually means to them.
- **Accountability must be clear and shared.** When an AI system makes a decision that affects a customer or the business, someone, a real human, needs to be responsible for that outcome. Not "the algorithm." Not "the data science team." A named leader who owns the results and looks the affected customer in the eye when something goes wrong.

- **Trust is built through transparency, not just performance.** An AI system can be highly accurate and still destroy trust if people don't understand how it works, can't challenge its decisions, or feel excluded from the process. When a loan officer can explain to a customer exactly why their application was declined, and what they can do about it, that's trust in action.
- **Employees are partners in change, not victims of it.** The organizations that succeed with AI are the ones that invest in their people (retraining, redesigning roles, communicating openly) rather than surprising them with automation. That means the operations analyst who's been with your firm for fifteen years sees AI as the best thing that happened to her career, not the end of it.
- **Governance is an enabler, not an obstacle.** Done well, AI governance doesn't slow you down. It gives you the confidence and organizational trust to move faster and scale further. It means a risk officer who feels empowered to say "yes, and here's how we do it safely" rather than reflexively saying no.

This philosophy is what I call "Human in the Lead." It doesn't mean a human approves every AI decision; that's impractical and often unnecessary. It means that humans set the intent, define the boundaries, and remain accountable for outcomes. AI operates within those boundaries, often with significant autonomy, but always within a framework of human purpose and human oversight.

Think of it like a jazz ensemble. The musicians are extraordinarily skilled, capable of playing with speed, precision, and improvisational range that continually surprises the listener. But they play within a structure: a key, a tempo, a shared understanding of where the song is headed. The bandleader doesn't play every note or dictate every riff. Instead, the bandleader sets the direction, calls the changes, and trusts each player to contribute something the others can't. The best moments happen when the players respond to each other in real time, the pianist hears something in the bass line and builds on it, the drummer shifts the feel, and everyone follows. That's "Human in the Lead." The AI surfaces patterns the human might miss; the human provides context, values, and judgment that the system cannot replicate. It's a relationship, not a hierarchy. And like any good ensemble, it works best when each player is clear about what they bring and trusts the others to do the same.

• • •

## A Taste of What's Inside

Before we go further, let me give you a glimpse of where this book will take you, because the ideas ahead are more surprising, more practical, and more immediately useful than any abstract summary can convey.

In Chapter 1, you'll learn why the history of technology in banking, from ATMs to mobile apps, consistently shows the opposite of what everyone predicted. ATMs didn't eliminate tellers; they made branches cheaper to run, so banks opened more of them, and the number of tellers actually grew for decades. Online banking didn't end branch visits; it shifted them from routine transactions to complex, high-value conversations. There is a pattern here, and it has direct implications for how AI will reshape roles in your organization, if you know how to read it.

In Chapter 2, you'll find a decision framework that distinguishes three models of human involvement in AI (Human-in-the-Loop, Human-in-the-Command, and Human-in-the-Lead) and a practical guide for choosing which model fits which use case. This isn't philosophy; it's a tool you can use in your next governance meeting.

In Chapter 3, you'll see how to use the Role Redesign Canvas to rethink jobs, not to eliminate them, but to make them more valuable. You'll see how relationship managers can evolve from data-gatherers to trusted advisors, how contact center agents can shift from script-followers to empathy specialists, and how operations staff can move from manual processors to exception analysts. These aren't aspirational labels. They're specific role redesigns with concrete examples of what changes and what stays.

In Chapter 6, you'll encounter a concept I call the "attention budget," the idea that human oversight is a finite resource, and that designing AI systems without accounting for the limits of human attention leads to rubber-stamp reviews that provide the illusion of oversight without the substance. You'll learn how to design intervention points that actually work, and why it matters whether oversight happens before or after a decision reaches the customer.

In Chapter 9, you'll read anonymized stories from real financial services AI projects: an operations transformation that nearly collapsed because the technology worked perfectly but the organization wasn't ready; a virtual assistant deployment that failed, was redesigned around human needs, and ultimately rebuilt customer trust; a fraud detection system that was highly accurate in aggregate but systematically unfair to specific customer segments; and a board that stumbled its way to effective AI governance after initial missteps. Each story follows a pattern you'll recognize, and each reveals lessons you can apply immediately.

And in Chapter 11, you'll find a People-Centric ROI Scorecard that goes far beyond traditional cost-savings math, because one of the most important lessons from the field is that what you measure shapes what your AI does. Organizations that measured only processing volume hollowed out their teams' expertise. Those that measured only containment rates trapped customers in loops with no human escape. The scorecard gives you a way to measure what actually matters: trust, employee engagement, fairness, and sustainable business value.

These are the kinds of ideas that run through every chapter, specific, grounded, and designed to change how you think and what you do. This book isn't a manifesto. It's a working toolkit for leaders who need to get AI right in an industry where the consequences of getting it wrong are measured in people's financial lives.

• • •

## Why Now? The Convergence of Three Forces

You might reasonably ask: hasn't AI been around in financial services for years? Why does it need a people-centric approach now?

The answer is that three forces are converging in ways that make the next few years uniquely consequential.

First, the technology itself has taken a qualitative leap. Generative AI and agentic AI aren't just faster or more accurate versions of what came before; they're fundamentally different. These systems don't just answer questions; they draft documents, make recommendations, take actions, and orchestrate complex workflows. In 2026, we're seeing AI agents that autonomously handle trade accounting, process know-your-customer checks, and route payment transactions, the last in under 200 milliseconds. This isn't a pilot anymore. It's operational reality, and it demands a governance approach that matches its capabilities.

Second, the regulatory environment is crystallizing rapidly. The EU AI Act, the world's first comprehensive AI regulation, is moving into full implementation, with credit scoring and lending decisions classified as "high-risk" AI systems that require human oversight, transparency, and auditability. (Fraud detection, notably, is expressly carved out of the Act's high-risk credit category, and anti-money-laundering monitoring remains governed by existing financial-crime regulation.) Regulators in the United States, the United Kingdom, and across Asia-Pacific are moving in the same direction. The window for figuring out governance after the fact is closing.

Third, and this is the force I believe matters most, the talent and trust landscape has shifted profoundly. Employees are anxious. Fear of job displacement is consistently the biggest barrier to AI adoption inside organizations. Talk to a mid-career relationship manager at almost any bank and you'll hear it: a quiet dread that the skills they've spent decades honing (reading a client's unspoken concerns, knowing when to push and when to listen, building trust across a kitchen table) are about to be rendered irrelevant by a machine. That fear is mostly unfounded, but it's real, and it's shaping behavior right now. Meanwhile, customers are skeptical; they want personalization but not surveillance, efficiency but not impersonality. A wealth management client told me recently that she didn't mind if AI helped her advisor prepare for their meeting, but she would leave the firm the day she felt like she was talking to a chatbot instead of a person. And boards are asking harder questions, often after reading about AI incidents at other institutions and wondering whether they're exposed.

These three forces (more powerful AI, tighter regulation, and heightened human stakes) mean that the old approach of "build the technology first, worry about the people later" isn't just ineffective. It's dangerous. The institutions that will thrive in this new environment are the ones that start with people and work outward.

• • •

## What This Book Will Give You

This book is a practical guide, not a philosophical treatise. It's written for financial services leaders (chief executives, business line heads, chief risk officers, chief technology officers, and the senior managers who work with them) who need to make real decisions about AI in the coming months and years.

Every chapter is designed to give you something you can use: a framework, a checklist, a decision model, or a way of thinking about a challenge you're likely facing right now. The ideas here are grounded in practice, drawn from real projects across banking, wealth management, and insurance, and pressure-tested against the regulatory expectations and organizational realities that define our industry.

There are many books about AI, and a growing number about AI ethics and governance. What makes this one different is its specificity. It is written exclusively for financial services, an industry with unique regulatory obligations, unique trust dynamics, and unique organizational complexity. It provides named, usable frameworks rather than abstract principles. And it is grounded in the real experiences of institutions that have tried, failed, learned, and tried again, not in laboratory experiments or theoretical ideals.

Here's how the book is organized:

**Part I: The Case for People-First AI** (this Introduction plus Chapters 1 and 2) makes the argument for why people-centric AI is essential, not optional, in financial services. Chapter 1 uses the surprising history of technology in banking, from ATMs to mobile, to reveal patterns that predict how AI will actually reshape the workforce. Chapter 2 introduces a three-part framework for human involvement (Human-in-the-Loop, Human-in-the-Command, and Human-in-the-Lead) with a decision model for choosing the right approach for each use case, and an accountability structure that ensures nobody gets to say, "That's not my problem."

**Part II: People—Employees, Customers, and Oversight** (Chapters 3 through 6) focuses on the people most directly affected by AI in finance. Chapter 3 provides the Role Redesign Canvas, a practical tool for transforming jobs so that relationship managers become trusted advisors instead of data-gatherers, and operations staff become exception analysts instead of manual processors. Chapter 4 tackles customer-facing AI with a design philosophy built on hyper-personalization with guardrails and a principle that every AI interaction must have a clear escalation path to a human. Chapters 5 and 6 get deeply practical about governance and human oversight, including the "attention budget" concept that explains why most oversight programs fail, and how to design intervention points that don't become rubber stamps.

**Part III: Building the Capability** (Chapters 7 through 9) addresses the organizational infrastructure you need. Chapter 7 provides tailored AI fluency programs for every level, from the board member who asks, “Can’t we just use ChatGPT for everything?” to the frontline analyst who needs to know when the model is wrong. Chapter 8 introduces a glass-box architecture principle (the idea that AI system internals should be visible to oversight functions, not locked inside the data science team) along with practical guidance on security, data minimization, and behind-the-firewall deployment. And while real-world stories are woven throughout every part of this book, Chapter 9 brings them together in four extended case vignettes: an operations transformation that nearly collapsed, a virtual assistant that rebuilt trust, a fraud system that was accurate but unfair, and a board that stumbled its way to effective governance.

**Part IV: Execution** (Chapters 10 through 12) is about doing the work. Chapter 10 provides a day-by-day 90-Day Roadmap for launching or resetting an AI initiative, with specific guidance on what to say to employees, customers, and regulators at each stage. Chapter 11 introduces the People-Centric ROI Scorecard, a measurement framework that goes beyond cost savings to capture trust, fairness, and employee engagement, because, as the case vignettes reveal, what you measure shapes what your AI becomes. And Chapter 12 looks ahead to the next wave, agentic AI, autonomous systems, and a Governance Readiness Framework for capabilities that don’t yet exist but soon will.

Each chapter is designed to stand on its own. If you’re a CRO grappling with governance, you can go straight to Chapters 5 and 6. If you’re a CHRO worried about workforce transformation, start with Chapter 3. If you’re a CEO who needs a 90-day plan, jump to Chapter 10. You don’t have to read sequentially, though I think you’ll find that the chapters build on each other in ways that deepen your understanding of the whole picture.

• • •

## A Word Before We Begin

I want to be honest about something: nobody has all the answers here. The technology is evolving faster than any of us can fully track. The regulatory landscape is still taking shape. And the organizational challenges around AI are genuinely hard; there's no playbook that works perfectly in every institution.

What I can offer is a way of thinking about these challenges that has proven effective across a range of real-world situations. It's a framework built on a simple insight: that the hardest parts of AI aren't the algorithms or the infrastructure. They're the human parts, the trust, the governance, the change management, the leadership. Get those right, and the technology has the best possible chance of delivering on its promise. Get them wrong, and no amount of technical sophistication will save you.

The institutions I've seen succeed with AI all share something in common. They didn't start with the question "What can AI do?" They started with a different question: "What do we want AI to do for our people (our employees, our customers, and the communities we serve), and how do we make sure humans remain in the lead as we get there?"

Back in that boardroom on the 32nd floor, the CEO didn't need a better algorithm. She needed a better answer to a fundamentally human question, one that would give every person at that table a shared sense of purpose, a clear role in what was coming, and the confidence that the institution they were building together would emerge from this transformation stronger, more trusted, and more deeply connected to the people it serves.

That's the question this book will help you answer.

Let's get started.

CHAPTER 1

# Why People-Centric AI Matters Now in Finance

Every few decades, a technology comes along that banking leaders are sure will change everything. And every time, the prediction is half right and half wrong.

When ATMs appeared in the late 1960s, the conventional wisdom was clear: branches would disappear and bank tellers would be out of work within a generation. What actually happened was more interesting. The number of bank tellers in the United States grew for decades after ATMs arrived. Why? Because ATMs made it cheaper to operate a branch, which meant banks opened more of them. And the tellers who remained shifted from counting cash to something more valuable: having conversations with customers, solving problems, and selling financial products. The technology didn't replace people. It changed what people did.

Online banking told a similar story. When the web arrived in the 1990s, some predicted the end of physical banking entirely. Customers would manage everything from their desktops. In reality, online banking did take over routine transactions (checking balances, paying bills, moving money), but it also made customers more aware of the products and services available to them. Branch visits dropped for simple tasks and increased for complex ones: mortgage applications, investment conversations, business lending discussions. Technology handled the routine; humans handled the meaningful.

Mobile banking repeated the pattern. The smartphone put the bank in everyone's pocket, and once again the prediction was that human interaction in banking was finished. And once again, what happened was subtler. Mobile made basic banking effortless, but it also raised customer expectations for speed, personalization, and responsiveness when they did need human help.

There's a lesson in this history that's directly relevant to what we're facing now with AI. In every previous technology wave, the institutions that thrived weren't the ones that used the technology to eliminate human interaction. They were the ones that used it to make human interaction more valuable. The technology succeeded not when it replaced people, but when it augmented them.

We're at one of those inflection points again. But this time, the stakes are higher, the technology is more powerful, and the margin for error is thinner. This chapter maps the landscape: the forces that make AI adoption unavoidable, the reasons why a people-centric approach isn't just nice to have but strategically essential, and the realities that make getting this right uniquely challenging in financial services.

• • •

## The Regulatory Ratchet Is Tightening

If you needed a single reason to take people-centric AI seriously, regulation would be it.

For most of the past decade, AI regulation in financial services existed primarily through the lens of existing frameworks: model risk management, fair lending rules, data protection laws. Regulators were watching, occasionally asking pointed questions, but there was no dedicated AI-specific regulatory regime. That era is over.

The EU AI Act, which entered into force in August 2024 and is rolling out in phases through 2027, is the world's first comprehensive AI regulation, and its implications for financial services are profound. The Act classifies AI systems by risk level, and many of the uses most relevant to banking fall squarely in the "high-risk" category: credit scoring and lending decisions, risk assessment and pricing in life and health insurance, and any AI used to evaluate the creditworthiness of individuals. Notably, the Act expressly exempts AI used to detect financial fraud from its high-risk credit category; fraud and anti-money-laundering systems remain governed primarily by existing financial-crime regulation. For the systems that do fall in the high-risk category, the Act requires human oversight, transparency, documentation, and ongoing monitoring. It requires that people affected by AI decisions can get meaningful explanations. It requires that the organizations deploying these systems can demonstrate they've tested for accuracy, robustness, and bias.

In November 2025, the European Banking Authority published a factsheet confirming that the AI Act doesn't replace existing banking regulation; it complements it. The practical implication? Banks don't get to treat AI governance as a separate, standalone initiative. They need to weave it into their existing risk management, compliance, and audit frameworks. That sounds manageable in theory. In practice, it means every AI system used in a regulated context needs to be inventoried, classified, documented, and subject to oversight mechanisms that can withstand regulatory scrutiny.

And Europe isn't alone. In the United States, the OCC, SEC, and Treasury Department have all intensified their focus on AI governance, with the Treasury's December 2024 report on AI in financial services laying down clear expectations around transparency and consumer protection, and state-level AI legislation proliferating. In the United Kingdom, the FCA and PRA are developing supervisory expectations focused on model governance and consumer outcomes. Singapore's MAS has been a leader in AI governance frameworks for finance. Australia's APRA is integrating AI risk into its broader operational resilience expectations. The details vary by jurisdiction, but the trajectory is unmistakable.

Across every major jurisdiction, regulators are converging on a risk-based approach to AI governance that puts human oversight, transparency, and accountability at the center. The institutions that have built these capabilities early will find compliance manageable. The ones that haven't will face an increasingly uncomfortable reckoning.

> *The window for figuring out AI governance after the fact is closing. Regulatory expectations are hardening into requirements, and the cost of being caught unprepared is rising with every enforcement cycle.*

• • •

## The Competitive Pressure Is Real, and Accelerating

While regulators tighten the guardrails, competitive pressure is pushing in the opposite direction: move faster, do more, spend more.

The numbers are striking. Broadridge's 2026 industry study found 80 percent of financial services firms reporting active AI use, up from 31 percent just a year earlier. Nearly every institution plans to increase or maintain its AI budget. And the nature of what's being deployed has changed dramatically. We're not talking about chatbots and simple automation anymore. We're talking about systems that can reason, learn, and act.

Consider what the industry leaders are doing. Goldman Sachs is developing autonomous AI agents, powered by Anthropic's Claude, to handle core trade accounting and client onboarding tasks. These aren't assistants that help humans do their work; they're digital co-workers that independently execute complex, multi-step processes. Lloyds Banking Group has declared 2026 the year of "enterprise-wide deployment" of agentic AI, expecting these systems to generate £100 million in value by automating fraud investigations and complex complaint handling.

And it's not just the global giants. Mid-sized banks, regional institutions, and wealth management firms are all racing to deploy AI across customer service, operations, compliance, and advisory functions. The same study found 26 percent of financial firms already using agentic AI, systems that don't just respond to queries but proactively take actions. Among those, more than half have moved beyond pilots into operational deployment.

There's a concept gaining traction in banking circles called the "10× bank," a vision (circulating among fintech leaders and increasingly echoed by traditional banking strategists) in which a single individual supported by a team of AI co-workers can deliver exponentially more output than a traditional team. The idea is that AI agents handle research, data analysis, compliance checks, and document preparation, while the human provides judgment, relationship management, and strategic direction. Whether or not that specific vision materializes in its most ambitious form, the underlying trend is clear: AI is compressing the gap between what large and small institutions can do. A well-equipped mid-sized bank with smart AI deployment can now compete with institutions ten times its size in certain capabilities. That's exhilarating if you're the mid-sized bank. It's terrifying if you're the large institution that's been relying on scale as its primary competitive advantage. We'll explore the implications of this model for workforce design in Chapter 3 and for architecture in Chapter 8.

The competitive pressure creates a genuine tension. Move too slowly, and you lose ground to more agile competitors. Move too fast without the right governance and people infrastructure, and you risk the kind of failure (regulatory, reputational, or operational) that can set you back years. Navigating this tension is one of the central leadership challenges of our moment. And it's why a people-centric approach matters so much: it's the framework that lets you move fast and get it right.

• • •

## "Move Fast and Break Things" Doesn't Work Here

Silicon Valley's most famous mantra was never a good fit for financial services. But in the age of AI, it's actively dangerous.

The reason is simple: in financial services, the things that break are people's lives. A flawed credit algorithm doesn't just produce a suboptimal outcome; it denies someone a home, a car, an education. A biased fraud detection model doesn't just create a false positive; it freezes someone's account while they're buying groceries for their family. A poorly designed robo-advisor doesn't just underperform; it erodes the retirement savings that someone spent decades building.

And the consequences extend beyond individual customers. In an industry built on trust, a single high-profile AI failure can ripple through an entire institution. Customers leave. Regulators investigate. The media reports. Your best people, the very ones you need to fix the problem, start looking elsewhere.

The data on AI readiness in financial services is sobering. A 2025 Infosys study of more than 1,500 senior executives found that only 2 percent of companies met its full standard for responsible-AI safeguards, and 95 percent of executives reported at least one AI incident inside their own organization. In financial services, where the consequences of AI failures are amplified by regulatory scrutiny and customer trust dynamics, these numbers should alarm every executive.

This doesn't mean we should be paralyzed. The answer to AI risk isn't to avoid AI; that's its own form of risk, as the competitive pressures we just discussed make clear. The answer is to deploy AI in ways that build in safeguards, human oversight, and accountability from the start, not as afterthoughts bolted on when something goes wrong.

Think about how the aviation industry approaches safety. Aviation is one of the most technology-intensive industries on earth. Modern aircraft are enormously complex systems with extensive automation. But the aviation industry doesn't treat safety and technology as opposing forces. Safety is embedded in every aspect of design, training, operation, and governance. Pilots are trained not just to fly but to manage automation, to know when to trust the systems, when to intervene, and how to make decisions under uncertainty. The result: commercial aviation is extraordinarily safe, despite, or rather because of, its sophistication.

Financial services needs a similar approach. We need to deploy AI aggressively and responsibly, treating governance not as a tax on innovation but as the foundation that makes innovation sustainable. We need leaders who understand that moving fast and getting it right are not contradictory goals; they're complementary ones, if you invest in the people and processes that make both possible.

• • •

## The Legacy Realities That Make This Harder

If the strategic case for people-centric AI is clear, the practical challenge is just as real. Financial services doesn't get to start from a blank slate. It starts from legacy systems, legacy culture, and legacy ways of working that create friction at every turn.

Start with the technology. Most banks run on core systems that are decades old, platforms built in COBOL, patched and extended over years, held together by integrations that would make an engineer weep. Data sits in silos: the retail bank has its data, the wealth management division has different data, operations has another set entirely. Getting a unified view of a single customer across these systems is often a multi-year project in itself. Now layer AI on top of that. AI systems are hungry for clean, connected, well-governed data. In most financial institutions, that's exactly what they don't have.

Then there's the talent challenge. Every financial institution in the world is competing for a limited pool of AI talent (data scientists, machine learning engineers, AI product managers) and competing not just with each other but with technology companies that can offer higher salaries, faster-moving cultures, and the allure of working on cutting-edge problems without regulatory constraints. But the talent gap isn't just about technical headcount. The hardest roles to fill are the hybrid ones: the model risk specialist who understands both machine learning and prudential regulation, the AI product manager who can translate between data science teams and compliance officers, the business leader who can evaluate an AI vendor's claims without being dazzled by a demo. That combination of technical and financial services fluency is exceptionally rare, and the institutions that can attract and retain it have a significant advantage, one that compounds over time as those people build the organizational muscle that technology alone cannot provide.

Perhaps the most underestimated challenge is cultural. Financial services organizations are, by design, risk-averse. That's not a flaw; it's a feature. When you're managing other people's money and operating under intensive regulatory oversight, a healthy respect for risk is exactly what you want. But risk aversion, when it becomes reflexive rather than thoughtful, can turn into innovation paralysis. Every AI initiative becomes a battle: the innovation team pushing forward, the risk team asking for more analysis, the compliance team requesting more documentation, and the business team losing patience with all of them.

The answer isn't to override these instincts. It's to channel them productively. A people-centric approach recognizes that risk teams and compliance teams aren't obstacles to AI adoption; they're essential partners in making it work. When risk professionals are involved early in AI design, they catch problems that would otherwise become expensive failures. When compliance teams help shape governance frameworks from the start, the path to regulatory approval is smoother, not slower. When frontline staff are consulted about how AI will change their work, adoption is faster, not harder.

The legacy realities are real. They make AI harder to deploy in financial services than in many other industries. But they're not insurmountable, and the institutions that learn to work with these realities rather than fight them will find they have a surprisingly strong foundation for sustainable AI adoption.

• • •

## People-Centric AI as a Competitive Differentiator

We've been talking about why people-centric AI is necessary, necessary for regulatory compliance, for managing risk, for navigating the legacy realities and organizational complexity that define our industry. But here's the part that makes this story genuinely exciting: it's not just a defensive play. It's a competitive advantage.

The evidence is starting to accumulate. A 2024 Wipro survey of US wealth management executives found that 70 percent of firms heavily leveraging AI reported a positive impact on client interactions, but with an important caveat: the positive impact occurs only when human advisors remain central to the client relationship. The firms that tried to use AI to disintermediate their advisors (to push clients toward fully digital, self-service experiences) didn't see the same results. The magic happened when AI made advisors better, not when it replaced them.

This pattern repeats across the industry. In operations, the banks that see the best results from AI automation are the ones that redesign roles around AI rather than simply eliminating headcount. In customer service, the virtual assistants that earn the highest satisfaction scores are the ones designed for seamless human handoff, not the ones optimized to keep customers trapped in a digital channel. In compliance, the most effective AI-driven monitoring systems are the ones that surface the right cases for human review, rather than drowning compliance officers in false positives.

There's a broader market dynamic at work here too. As AI becomes commoditized, as the same foundational models become available to every institution, the differentiator shifts. It's no longer about who has the best algorithm. The technology becomes table stakes. What isn't commoditized, and what can't be, is how an institution earns trust, engages its people, and serves its customers. Those are the competitive moats that matter now, and they're fundamentally human.

> *In a world where every bank has access to the same AI models, the competitive advantage belongs to the institutions that deploy those models in ways that deepen trust, empower employees, and serve customers better. That's a people advantage, not a technology advantage.*

• • •

## A Framework: The AI Readiness Landscape

Before we move on, I want to offer a simple framework that I've found useful for helping leadership teams understand where they stand and what they need to prioritize.

Think about your institution along two dimensions: regulatory pressure and organizational readiness. Regulatory pressure is about how exposed you are to AI-specific regulation, how many of your AI use cases fall into high-risk categories, how active your regulators are in this space, and how close you are to regulatory deadlines. Organizational readiness is about your internal capability, your data infrastructure, your governance maturity, your talent, and your cultural readiness for AI-augmented work.

| | Low Organizational Readiness | High Organizational Readiness |
|---|---|---|
| High Regulatory Pressure | URGENT ACTION<br>Focus on governance foundations first. Inventory your AI systems, establish oversight structures, and close the most critical compliance gaps before scaling further. | SCALE WITH CONFIDENCE<br>You're well-positioned. Focus on expanding use cases, deepening governance maturity, and building the people-centric metrics that demonstrate value to regulators and stakeholders. |
| Low Regulatory Pressure | BUILD FOUNDATIONS<br>Use the breathing room wisely. Invest in data infrastructure, talent, and cultural readiness now, before regulatory pressure catches up, because it will. | **ACCELERATE STRATEGICALLY**<br>You have capability and room to experiment. Push ahead on high-value use cases while proactively building the governance framework that will be expected as regulation evolves. |

*Figure 1.1: The AI Readiness Landscape*

Most financial institutions I work with find themselves in the top-left quadrant: high regulatory pressure and low organizational readiness. That's an uncomfortable place to be, but it's also a clarifying one. It tells you exactly what to prioritize: governance foundations, leadership alignment, and a realistic assessment of your current AI portfolio. Not more pilots. Not another proof-of-concept. The foundational work that lets everything else succeed.

Wherever you sit on this landscape, the path forward is the same in principle, even if the pace and emphasis differ: build the people infrastructure (governance, fluency, change management, measurement) that turns AI ambition into AI reality.

• • •

## Where We Go from Here

This chapter has laid out the landscape: a regulatory environment that's hardening, competitive pressure that's accelerating, legacy realities that create genuine friction, and a growing body of evidence that people-centric approaches are the ones that actually work.

If there's a single takeaway, it's this: the institutions that will succeed with AI in financial services are not the ones with the biggest budgets or the most advanced algorithms. They're the ones that understand the most important variable is human. Human trust. Human judgment. Human accountability. Human leadership.

In the next chapter, we'll translate this conviction into a set of actionable principles. We'll define what "Human in the Lead" looks like in practice, the different models of human involvement, the accountability structures, and the non-negotiable design principles that make people-centric AI more than a philosophy. We'll give you the conceptual toolkit you need to start making different decisions about AI in your organization.

Because understanding why this matters is the first step. Knowing how to do it is what changes the outcome.

CHAPTER 2

# Principles of People-Centric AI in Financial Services

Let me tell you about two banks.

The first, call it Bank A, built a credit decisioning AI that was, by any technical measure, impressive. It processed loan applications in seconds rather than days. Its predictive accuracy was superior to the old scorecard model. It reduced default rates. On the dashboard, everything looked green.

But on the ground, the story was different. Customers who were declined couldn't get a meaningful explanation. The best the call center could offer was "the system made the decision based on your profile." Branch staff, who used to exercise judgment on borderline cases (factoring in a customer's long relationship with the bank, or a temporary dip in income they knew was about to recover) found they had no way to override or even question the AI's output. The compliance team couldn't demonstrate to the regulator how the model handled protected characteristics. Within a year, customer complaints about lending decisions had tripled. A regulatory inquiry was opened. The model was quietly rolled back.

Now consider Bank B. Their credit AI was less sophisticated, a bit slower, a bit less "optimized" on paper. But it was designed differently. Advisors could see the key factors driving each recommendation. They could override the AI when they had information the model didn't, and the system logged those overrides for learning and audit purposes. Customers who were declined received a clear explanation of the primary reasons and a pathway to reconsideration. The compliance team had full documentation of the model's logic, its fairness testing results, and its ongoing performance metrics.

Bank B's model had lower raw predictive accuracy than Bank A's. But it had higher adoption by staff, fewer customer complaints, smoother regulatory reviews, and, here's the kicker, better business outcomes, because the human overrides caught edge cases the model missed and the transparency built the kind of trust that keeps customers loyal.

These two banks illustrate the core argument of this chapter: in financial services, how you deploy AI matters as much as what you deploy. The principles you design around, or fail to design around, determine whether AI strengthens your institution or quietly undermines it.

This chapter defines those principles. We'll start with the most fundamental question, namely how humans should be involved in AI-driven decisions, and work outward through accountability, augmentation, and the non-negotiable design commitments that separate trustworthy AI from everything else.

• • •

## Three Models of Human Involvement

One of the most common phrases in AI governance is "human in the loop." It shows up in regulations, in corporate policies, in conference keynotes. And while it's a useful starting point, it's also dangerously simplistic.

In practice, there isn't just one way to keep humans involved in AI decisions. There are at least three distinct models, and choosing the right one for each use case is one of the most consequential decisions a financial services leader can make. Get it wrong, and you either create expensive bottlenecks that negate

AI's value, or you create the illusion of oversight where none meaningfully exists.

Here's how I think about it.

## Human-in-the-Loop

In this model, a human reviews and approves every AI output before it takes effect. The AI recommends; the human decides. Nothing happens without explicit human sign-off.

This is the right model for high-stakes, low-volume decisions where the consequences of getting it wrong are severe and where human judgment adds genuine value. Think of a large commercial loan approval. The AI can synthesize financials, assess risk indicators, and flag anomalies far faster than a human analyst. But the final decision (weighing the relationship, the strategic value of the client, the nuances that don't show up in the data) benefits enormously from an experienced human underwriter.

The danger of Human-in-the-Loop is scale. If you apply it to high-volume decisions, you create a bottleneck that defeats the purpose of using AI in the first place. And there's a subtler risk: automation bias. When humans review hundreds of AI recommendations and agree with most of them, they can slip into rubber-stamping, approving outputs without genuine scrutiny. At that point, you have the cost of human involvement without the benefit of human judgment.

## Human-in-the-Command

In this model, the human doesn't review every individual decision. Instead, the human sets the rules, defines the parameters, monitors aggregate performance, and intervenes on exceptions. The AI operates within the boundaries the human has defined and escalates cases that fall outside those boundaries.

This is the right model for medium-stakes, high-volume decisions where individual review isn't practical but meaningful oversight is still essential. Fraud alert triage is a good example. An AI system might process millions of transactions per day and flag several thousand for potential fraud. A human can't review every flag. But a human can define the escalation thresholds, review the most ambiguous or high-value cases, monitor false positive and false negative rates, and adjust the system's parameters based on what they observe.

The key to making Human-in-the-Command work is the quality of the monitoring and exception-handling mechanisms. The human needs to see the right information at the right time: aggregate performance metrics, trend data, representative samples of decisions, and, critically, the cases where the AI is least confident or where its decisions could have the most impact.

## Human-in-the-Lead

This is the model at the heart of this book, and it's the most misunderstood.

Human-in-the-Lead means a human defines the strategic intent, the ethical boundaries, and the accountability framework within which AI operates. The AI may act with significant autonomy (processing transactions, generating personalized recommendations, routing payments), but the human remains ultimately accountable for outcomes. The human doesn't review every decision or even every exception. Instead, the human ensures that the system itself is designed, governed, and monitored in ways that produce trustworthy results.

Think of it this way. In Human-in-the-Loop, the human is the pilot, hands on the controls for every maneuver. In Human-in-the-Command, the human is the air traffic controller, monitoring the airspace, setting routes, and intervening when something goes wrong. In Human-in-the-Lead, the human is the aviation regulator and airline CEO combined, designing the safety framework, setting the standards, ensuring the systems are built and maintained to those standards, and taking accountability when things go wrong.

Human-in-the-Lead is the right model for scaled, lower-risk operations where the volume of decisions makes individual or even exception-based review impractical, but where robust governance, monitoring, and accountability structures ensure that the AI operates within acceptable boundaries. Personalized marketing, routine transaction processing, and standard account servicing are typical examples.

> *"Human-in-the-Lead" is not the same as "Human-in-the-Loop." It's not about reviewing every decision. It's about ensuring that humans set the purpose, define the boundaries, and remain accountable for outcomes, even when AI operates with significant autonomy.*

• • •

## Choosing the Right Model: A Decision Framework

The choice between these three models isn't arbitrary. It depends on four factors (decision stakes, decision volume, reversibility, and regulatory sensitivity) that vary from one use case to the next. The matrix below maps these factors to each model:

| Factor | Human-in-the-Loop | Human-in-the-Command | Human-in-the-Lead |
|---|---|---|---|
| **Decision Stakes** | Very high, life-altering or irreversible consequences | Medium, significant but manageable consequences | Lower, limited individual impact, aggregate patterns matter more |
| **Volume** | Low, tens to hundreds per day | High, thousands to millions per day | Very high, continuous, real-time processing |
| **Reversibility** | Low, difficult or impossible to undo | Medium, correctable with effort and time | High, easily reversed or adjusted |

| Factor | Human-in-the-Loop | Human-in-the-Command | Human-in-the-Lead |
|---|---|---|---|
| **Regulatory Sensitivity** | High, explicit requirements for human review of individual decisions | Medium, requirements for monitoring, auditability, and exception handling | Lower, general governance and accountability expectations |
| **Typical Examples** | *Large commercial loans, complex investment advice, major credit decisions* | *Fraud alert triage, AML monitoring, customer complaint routing* | *Personalized marketing, routine transactions, standard account servicing* |

*Figure 2.1: The Human-in-the-Lead Decision Matrix*

A few important caveats. First, these models aren't static. As an AI system matures, gains a performance track record, and demonstrates reliability, it may be appropriate to shift from Human-in-the-Loop to Human-in-the-Command, or from Human-in-the-Command to Human-in-the-Lead. The movement should always be earned through evidence, not assumed through ambition.

Second, the models can coexist within a single system. A credit decisioning AI might operate in Human-in-the-Lead mode for standard, low-risk applications, escalate to Human-in-the-Command for cases above a certain threshold, and require full Human-in-the-Loop approval for the largest or most unusual applications. This layered approach is often the most practical and effective.

Third, and this is critical, choosing Human-in-the-Lead does not mean choosing less oversight. It means choosing a different kind of oversight: one focused on system design, governance, monitoring, and accountability rather than individual decision review. Done well, it can be more rigorous than Human-in-the-Loop, because it catches systemic issues that individual review would miss.

• • •

## Shared Accountability: Nobody Gets to Say "That's Not My Problem"

Here's a scenario I've witnessed more times than I'd like. An AI model produces a biased outcome, say, disproportionately declining loan applications from a particular demographic group. The question is asked: who is responsible?

The business team says, "We just requested the model. Technology built it." The technology team says, "We built it to the specifications we were given. The data came from the business." The risk team says, "We weren't consulted until after deployment." And the compliance team says, "We flagged concerns in our review, but we were told the timeline couldn't accommodate a full assessment."

Everyone has a reasonable excuse. And nobody is accountable.

This is the accountability gap, and it's one of the most dangerous dynamics in AI deployment. It emerges whenever responsibility for AI outcomes is vague, diffused, or assumed to belong to someone else. And it's endemic in financial services, where organizational structures tend to be siloed and where the complexity of AI systems makes it easy for each function to point at another.

People-centric AI demands clear, shared accountability. Not shared in the sense that everyone is a little bit responsible, which in practice means no one is, but shared in the sense that each function has a defined and non-delegable role:

- **Business** owns the intent. They define what the AI is for, what outcomes it should produce, and what customer or business value it should deliver. If the AI is solving the wrong problem or optimizing for the wrong metric, that's a business failure.
- **Technology** owns the implementation. They build, test, and deploy the AI system to the specifications defined by the business and within the constraints defined by risk and compliance. If the model is technically flawed, poorly

engineered, or inadequately tested, that's a technology failure.

- **Risk** owns the controls. They define the risk appetite, review the model's risk profile, and establish the monitoring and escalation mechanisms. If risks materialize that should have been anticipated and mitigated, that's a risk failure.
- **Compliance** owns the assurance. They verify that the AI system meets regulatory requirements, internal policies, and ethical standards. If the system violates regulations that it should have been designed to comply with, that's a compliance failure.

Notice the pattern: each function can fail independently, and each is accountable for its own domain. But, and this is the crucial part, none of them can succeed independently. A technically brilliant model that solves the wrong problem fails. A well-intentioned initiative that's poorly engineered fails. A compliant system fails if no one uses it because the controls are too burdensome. Success requires all four functions working together, with clear lines of responsibility and regular, structured communication.

| AI Lifecycle Phase | Business | Technology | Risk | Compliance |
|---|---|---|---|---|
| **Use case definition & intent** | A | C | C | I |
| **Data selection & preparation** | C | A | R | C |
| **Model design & development** | C | A | R | C |
| **Fairness & bias testing** | I | R | A | R |

| AI Lifecycle Phase | Business | Technology | Risk | Compliance |
|---|---|---|---|---|
| Pre-deployment validation | C | R | A | R |
| Deployment & go-live | R | A | R | R |
| Ongoing monitoring & drift | I | R | A | C |
| Incident response | R | R | A | R |
| Periodic review & recalibration | C | R | A | R |

**A = Accountable; R = Responsible; C = Consulted; I = Informed**

*Figure 2.2: AI Accountability RACI. A Starter Template*

I want to emphasize the word "starter." This template is a conversation tool, not a compliance document. Every institution will need to adapt it to its own structure, culture, and regulatory context. The point is not to follow this exact matrix but to have the conversation it forces: Who is accountable for each phase of the AI lifecycle? Does everyone agree? Are there gaps? The institutions that have this conversation early, and revisit it regularly, are the ones that avoid the finger-pointing when something goes wrong.

• • •

## Augmentation Over Replacement

There's a persistent narrative around AI that goes something like this: AI will do the work that humans currently do, but faster, cheaper, and more accurately. The implication is that the goal of AI is to replace human labor. And in financial services, where labor costs are significant and many processes are repetitive, this narrative has real appeal in boardrooms.

It's also wrong. Or rather, it's wrong as a strategy, even when it's sometimes right as a description of what's technically possible.

The problem with the replacement mindset isn't that AI can't automate tasks. It clearly can, and it should, when automation improves outcomes. The problem is that replacement as a primary strategy misses the larger opportunity and creates the very resistance that causes AI initiatives to fail.

Consider a wealth management advisor. Could AI generate a portfolio recommendation? Technically, yes. Could it produce a financial plan? Probably. Could it have the conversation with a recently widowed client who is overwhelmed, grieving, and terrified about her financial future, a conversation that requires empathy, patience, sensitivity, and the ability to read what someone isn't saying? No. And that conversation is where the real value of the advisory relationship lives.

The augmentation approach says: let AI handle what AI does best (data synthesis, pattern recognition, scenario modeling, document preparation, compliance checking) so that the advisor can spend more time on what humans do best: building relationships, exercising judgment, providing empathy, and navigating the kind of complexity that doesn't reduce to an algorithm.

This isn't a soft argument. It's a hard-nosed business argument. The firms that augment their people rather than replace them see:

- **Higher adoption**: When employees see AI as a tool that makes their work better, they embrace it. When they see it as a threat, they resist, and resistance can be subtle, persistent, and devastating to an AI initiative's success.
- **Better outcomes**: Human-AI partnerships catch errors that neither humans nor AI would catch alone. The human spots the edge case the model misses; the AI surfaces the pattern the human overlooked.
- **Stronger customer relationships**: Customers in financial services want human connection for the moments

that matter most. AI that frees advisors to provide that connection deepens loyalty. AI that removes it destroys it.

- **More resilient organizations**: An institution that has developed its people's skills alongside its AI capabilities can adapt when the technology changes, when regulations shift, or when something goes wrong. An institution that has simply replaced people with AI is brittle.

There are, of course, situations where full automation is genuinely the right answer. We'll take that question head-on before this chapter closes.

The question is never "Can AI do this instead of a human?" The question is "How should AI and humans work together to produce the best possible outcome?" That reframing changes everything: the design, the governance, the change management, and, ultimately, the results.

• • •

## The Non-Negotiables: Transparency, Explainability, Fairness, and Contestability

Every financial institution claims to value transparency, fairness, and accountability. But when these values meet the pressures of AI deployment (speed to market, competitive advantage, cost reduction), they often become aspirational rather than operational. "We'll add explainability in the next version." "Fairness testing is on the roadmap." "We'll build the customer appeal mechanism once the system is stable."

People-centric AI doesn't accept these deferrals. It treats four design principles as non-negotiable, meaning they are built into AI systems from day one, not retrofitted after deployment.

## Transparency

Transparency means that the relevant stakeholders can see what the AI system does and why. This doesn't mean publishing the source code or exposing proprietary algorithms. It means that the people who need to understand the system (the business owners, the oversight functions, the regulators, and, where appropriate, the customers) can get answers to basic questions: What data does this system use? What is it optimizing for? How does it make decisions? What are its known limitations?

Transparency is the foundation on which all other principles rest. You can't hold an AI system accountable if you can't see what it's doing. You can't test it for fairness if you don't understand its logic. You can't design meaningful oversight if the system is a black box.

## Explainability

Explainability is related to transparency but distinct. Transparency is about making information available; explainability is about making it understandable. An AI system might be transparent in the sense that all its documentation is accessible, but not explainable if that documentation is so technical that the people who rely on the system can't make sense of it.

In financial services, explainability needs to work at multiple levels. A risk analyst needs to understand the model's statistical properties and performance characteristics. A relationship manager needs to understand why the AI recommended a particular product for a particular client. A customer needs to understand why their application was declined. These are fundamentally different explanations, and a well-designed AI system provides all of them, tailored to each audience's context and expertise.

## Fairness

AI systems learn from historical data, and historical data encodes historical patterns, including patterns of discrimination and inequality. A credit model trained on decades of lending data may learn to replicate the biases embedded in that data, effectively automating and scaling discrimination that human underwriters might have recognized and corrected.

Fairness in AI isn't about achieving some abstract ideal of perfect equality. It's about actively testing for differential treatment across protected characteristics (race, gender, age, disability, geography) and making deliberate choices about how to address disparities when they're found. This means fairness testing before deployment, ongoing monitoring after deployment, and clear processes for what happens when unfair patterns are detected.

Fairness is also about vulnerable customers, people who are financially stressed, elderly, digitally excluded, or otherwise at risk of being harmed by AI systems that weren't designed with their needs in mind. A people-centric approach asks: does this AI system work well for everyone, including the people who are most vulnerable? If the answer is no, deployment waits until it does.

## Contestability

This is the principle that gets the least attention and may be the most important for maintaining trust.

Contestability means that the people affected by AI decisions, customers and employees, have meaningful ways to challenge those decisions. Not just a "Contact Us" form that disappears into a queue. Not just a generic explanation letter generated by the same system that made the decision. A genuine pathway to have the decision reviewed by a human who has the authority, the information, and the willingness to override the AI when it got it wrong.

Contestability matters for two reasons. First, it's a safety net. No AI system is perfect, and contestability ensures that errors are caught and corrected. Second, and more profoundly, it's a signal of respect. It tells customers and employees that they are not subordinate to the algorithm, that their experience, their knowledge, and their judgment matter. In an industry built on relationships, that signal is invaluable.

Consider what this looks like in practice. A small-business owner applies for a line of credit and is declined by the bank's AI-driven underwriting model. Under a contestability framework, she doesn't just receive a form letter. She gets a clear explanation of the factors that drove the decision, a named contact who can review her case, and a defined timeline for resolution. When the reviewer examines her file, he discovers that the model weighted a brief industry downturn too heavily and overlooked three years of steady revenue growth. The decision is reversed, the credit line is approved, and the model's weighting is flagged for recalibration. Without contestability, that owner walks away, and the bank never learns its model has a blind spot.

• • •

## The Strongest Objection: "While We Govern, They Ship"

Every argument in this chapter eventually meets the same objection. It usually comes from someone whose judgment I respect, a CEO with a board breathing down her neck, a head of digital with a P&L to defend. It goes something like this:

*"Everything you've described costs time and money. Committees, testing, documentation, appeal pathways: each one is friction. Meanwhile, our competitor is shipping. They put a copilot in front of every banker last quarter. Their chatbot already handles half their service volume. If they capture our customers while we're perfecting our governance, the most elegant accountability framework in the world won't save us. Markets punish slow more reliably than they punish sloppy."*

I take this objection seriously, and you should too, because it isn't wrong about everything.

It's right that speed matters. AI capability is compounding, customer expectations are resetting annually, and there is no prize for the most thoroughly documented program that never launched. An institution that treats governance as a reason to defer every decision is not being careful; it's being absent.

It's also right that over-governance is real. I've seen risk functions where "no" is the default answer to anything unfamiliar, committees that exist mainly to diffuse responsibility, and review cycles that run longer than the builds they review. If your AI governance produces neither speed nor safety, only documentation, then the objection wins, and it deserves to.

And it's right that not every AI decision warrants the same scrutiny. Treating a marketing subject line like a mortgage decision isn't prudence. It's a failure to think.

But notice what the objection quietly assumes: that governance and speed trade off against each other, that every hour spent on boundaries is an hour not spent shipping. That assumption doesn't survive contact with the evidence, or with the experience of anyone who has cleaned up after a fast deployment.

Start with the evidence. Nearly every financial firm says it's racing to deploy AI agents; barely one in ten has done so at scale. The bottleneck isn't ambition, and it isn't velocity. It's that most institutions lack the structures that make velocity safe to sustain, so their pilots stall in review, their front line doesn't trust the output, and their lawyers won't sign off on scale. Speed without governance doesn't actually produce speed. It produces stuck.

Then look at what happens to the ones that do ship fast without the foundations. Bank A, from the start of this chapter, shipped faster than Bank B. Where is Bank A's model now? Rolled back, under regulatory inquiry, its rework costing multiples of what the governance would have, and that's before counting the customers who left and won't say why. The competitor who ships without controls hasn't escaped the cost of governance.

They've borrowed against it. Incidents, findings, and trust failures are the interest payments, and they compound.

Here's the part the objection misses entirely: done well, governance is what speed is made of. Watch a Formula 1 pit stop. It is one of the fastest coordinated actions in sports, and it is fast precisely because every motion, every role, and every failure mode has been defined in advance. Nobody in the pit lane is improvising, and nobody is waiting for permission. Clear trust boundaries do the same thing for AI teams: when a product team knows exactly which decisions the system may make on its own, which need review, and which are out of bounds, they don't queue outside committee rooms. They build. The boundaries are the speed.

So when a leader tells me their less-careful competitor is moving faster, my answer is: maybe, for now, and on borrowed time. The real choice was never speed versus safety. It's earned speed, the kind your boundaries, your evidence, and your people can sustain, versus borrowed speed, which someone always pays back. Usually at the worst possible moment, in front of a regulator.

## When Full Automation Is the Right Answer

The honest version of this argument has to concede something in return. People-centric AI does not mean a human reviewing everything, and there are real cases where inserting a human into the flow adds no judgment, only latency. Sometimes the latency itself is the risk. A payment routed in 200 milliseconds cannot wait for a reviewer. A fraud block that needs human approval before it takes effect is a fraud that has already happened. A document classifier handling two million pages a month gains nothing from a human skimming page 1,374,206.

Full automation is the right answer when four conditions hold together. The stakes of any individual decision are low. Errors are reversible and correctable in the aggregate. The system has earned an evidence base, a track record under monitoring, not a vendor's benchmark. And the institution maintains genuine, tested intervention capability: live monitoring, defined escalation, and a kill switch someone actually knows how to use. Notice that this is simply Human-in-the-Lead from Figure 2.1 doing its job. The human hasn't left; she has moved from reviewing decisions to owning the system that makes them.

The converse failure deserves equal honesty. Keeping humans in review loops where they add no signal doesn't create safety; it creates automation bias, rubber stamps, and what a later chapter will call oversight theater. A reviewer confirming two hundred AI decisions a day isn't oversight; he's a liability shield, and a thin one. It's worse than honest automation, because it diffuses accountability while adding cost and delay.

So the discipline this book argues for is not "slower everywhere." It's matching the oversight model to the decision, and then moving exactly as fast as the boundaries allow. In practice, that means some of your systems should get more autonomy than cautious institutions allow today, and others far less than the vendors are selling. Knowing which is which: that's the job.

• • •

## Principles in Practice

Let me bring this back to where we started: the tale of two banks.

Bank A had a technically superior model. But it was designed without transparency (staff couldn't explain decisions), without meaningful explainability (customers couldn't understand denials), without adequate fairness testing (the regulator found gaps), and without contestability (there was no genuine appeal mechanism). The principles weren't there, and the system failed, not technically, but in every other way that matters.

Bank B designed around the principles. Not because they were legally required to, though increasingly they are, but because they understood that in financial services, trust is the product. They chose a slightly less "optimized" system that their people could explain, their customers could understand, and their regulators could examine. And it outperformed.

That's the case for principles. Not as abstract ideals that look good in a corporate values statement, but as practical design decisions that determine whether AI builds trust or erodes it.

In the next chapter, we turn from principles to people. Specifically, to the employees whose roles, skills, and daily work AI is reshaping right now. How do we redesign work in a way that's honest about what's changing while being intentional about what matters? That's where we're headed.

CHAPTER 3

# AI and the Future of Work in Banks and Wealth

Sarah has been a relationship manager at a mid-sized regional bank for eleven years. She's good at her job, consistently in the top quartile for client retention and cross-sell revenue. Her clients trust her. Her branch manager relies on her. She knows the business.

A year ago, Sarah spent roughly 60 percent of her time on tasks that had little to do with the thing she's actually good at: building relationships with clients. She was gathering data from three different systems to prepare for client reviews. She was manually assembling compliance documentation. She was writing meeting notes into the CRM, formatting reports, and chasing down product information that should have been at her fingertips but wasn't.

Then her bank rolled out an AI platform, slowly at first, a document summarizer here, an automated report generator there. Then more aggressively: an AI assistant that could pull together a client's full financial picture across accounts, generate a draft meeting brief, flag relevant life events from recent interactions, and pre-populate compliance forms.

On paper, it was a triumph. Sarah's administrative burden dropped dramatically. She had more time for client conversations. The technology worked.

But Sarah didn't feel triumphant. She felt disoriented.

With 60 percent of her previous workload automated, she was supposed to spend more time on "high-value activities." But nobody had clearly defined what those were. Her performance metrics still emphasized the same things they always had: number of client meetings, cross-sell ratios, compliance completion rates. Her training hadn't changed. Her job description hadn't been updated. In team meetings, managers talked about AI as if it were just a faster version of the old tools, when in reality it was fundamentally changing what the job required.

Sarah's story is not unusual. It's the norm. Across the financial services industry, AI is automating tasks at an unprecedented pace. But the harder work (redesigning the roles, retraining the people, redefining what "good" looks like in an AI-augmented world) is lagging far behind the technology. That gap is where anxiety, disengagement, and failed AI initiatives live.

This chapter is about closing that gap.

• • •

## The Shift: From Tools to Teammates

To understand why this moment feels different, we need to be clear about what's changed.

For most of the last two decades, technology in banking was a tool. A tool you logged into, a tool you queried, a tool that did a specific thing when you pressed a specific button. The CRM was a tool. The risk calculator was a tool. Even the first-generation chatbots were tools, programmatic, rule-based, limited to the scripts they were given.

What's emerging now is fundamentally different. Generative AI and agentic AI don't just respond to specific commands. They understand context. They synthesize information across sources. They draft, recommend, and, increasingly, take actions. An AI agent in a bank today might autonomously prepare a client brief by pulling data from the CRM, the portfolio management system, and recent market research. It might draft a follow-up email after a client meeting, personalized to the conversation that just happened. It might process a know-your-customer (KYC) check end-to-end, flagging only the cases that need human judgment.

The numbers tell the story. As we saw in Chapter 1, about a quarter of financial firms are already using agentic AI, and most of those have moved beyond pilots into operational deployment. Goldman Sachs is building autonomous agents for core operations work. Accenture is helping banks compress KYC work that once consumed operations teams for weeks into workflows measured in days and hours. Payment routing systems are making optimization decisions in fractions of a second, faster than any human could process the inputs, let alone make the call.

The language is shifting to match. People increasingly talk about AI as a "digital co-worker" or "teammate" rather than a "tool." And that language shift matters, because it changes what we expect from the AI, and from the humans who work alongside it.

When AI was a tool, the human's job was to use it. When AI is a teammate, the human's job is to work with it, to know what it's good at, to understand its limitations, to supervise its outputs, and to do the things it can't. That's a different job. And in most financial institutions, it's a job that hasn't been designed yet.

• • •

## Redesigning Roles, Not Just Automating Tasks

Here's a trap that's easy to fall into: you identify the tasks that AI can automate, you automate them, and you assume the role will take care of itself. The remaining tasks will expand to fill the time. People will figure it out.

They won't. Or rather, some will and many won't, and the ones who struggle won't struggle because they're not talented or hardworking. They'll struggle because the role they were hired for, trained for, and evaluated on no longer exists in its original form, and nobody has articulated what the new role looks like.

Intentional role redesign is one of the most important and most neglected steps in people-centric AI. Here's what it looks like for four key roles in banking and wealth management.

## Relationship Managers: From Data-Gatherers to Trusted Advisors

Sarah's story captures this transition. When AI handles data gathering, report preparation, and compliance documentation, the relationship manager's core value shifts decisively toward the human skills: listening deeply to understand a client's real concerns (not just their stated ones), synthesizing complex information into clear advice, navigating emotional conversations about money and life transitions, and building the kind of trust that keeps clients loyal across decades.

This sounds obvious, but the implications are profound. It means relationship managers need training in consultative selling, empathetic communication, and complex financial planning, not just product knowledge and system navigation. It means performance metrics need to shift from activity counts (number of meetings, number of calls) to outcome measures (client satisfaction, depth of relationship, quality of advice). And it means managers need to coach differently, less "did you make your numbers?" and more "how are you using the time AI has freed up to deepen your client relationships?"

## Contact Center Agents: From Script-Followers to Empathy Specialists

AI virtual assistants are increasingly handling routine customer queries: balance inquiries, transaction status, password resets, basic product information. The best implementations I've seen resolve roughly half of inbound contacts without human involvement.

What does that leave for the human agents? The hard stuff. The calls that are complex, emotional, or both. The customer who's been the victim of fraud and is frightened. The elderly client who can't navigate the app and feels abandoned by their bank. The small business owner who needs help understanding their cash flow during a difficult quarter. These are situations that require empathy, patience, creative problem-solving, and the ability to exercise judgment, exactly the things AI can't do well.

The redesigned contact center role is not a lesser version of the old role with the easy calls removed. It's a harder, more demanding, and more valuable role. It requires emotional intelligence, deep product knowledge, and the authority to make decisions on the spot, to waive a fee, to escalate a case, to arrange a callback from a specialist. Treating these agents like script-followers, which is how many contact centers still operate, is a recipe for burnout and poor customer outcomes.

## Operations Staff: From Manual Processors to Exception Analysts

In operations, AI is automating the routine, high-volume work: processing transactions, reconciling accounts, generating standard reports, handling routine exceptions. What remains for humans is the non-routine: the exceptions that the AI can't classify, the discrepancies that don't match any pattern, the cases that require investigation and judgment.

This is a significant shift. The old operations role was largely about speed and accuracy in repetitive tasks. The new role is about analytical thinking, problem-solving, and judgment. An operations analyst in an AI-augmented environment needs to understand how the AI makes decisions well enough to spot when it's wrong. They need to be comfortable with ambiguity, because the cases that reach them are, by definition, the ones the AI couldn't resolve. And they need to contribute to the AI's improvement, flagging errors, suggesting new rules, helping train the system on edge cases.

There's also a subtle workforce planning challenge here. In many operations teams, the routine cases served as a training ground for junior staff. If AI handles the routine work, how do new analysts develop the judgment needed to handle exceptions? This is a solvable problem, but only if you design for it deliberately, through structured apprenticeship programs, supervised case review, and simulation exercises.

### Risk and Compliance Professionals: From Manual Reviewers to AI Oversight Specialists

Perhaps the most consequential role transformation is in risk and compliance. As AI takes on more of the analytical and monitoring work (running risk models, scanning for compliance issues, flagging suspicious activity), the role of the risk professional shifts from doing the analysis to overseeing the system that does the analysis.

This is the role Chapter 6 will explore in depth, but it's worth flagging here because it requires a fundamentally different skill set. AI oversight specialists need to understand both the domain (credit risk, market risk, regulatory compliance) and the technology (how the models work, what their limitations are, what "drift" looks like). They need the confidence to challenge the AI's outputs, to say "this doesn't look right" even when the model's track record is strong. And they need institutional support for that challenge function: a culture where questioning the AI is valued, not penalized.

• • •

## The Role Redesign Canvas

When I work with leadership teams on role redesign, I use a simple five-column canvas that forces the right conversations. Here's how it works.

| Current Tasks | AI-Augmentable | New Human-Value Tasks | Skills Gap | Training Plan |
|---|---|---|---|---|
| *List every task the role currently performs, with approximate time allocation* | *Which tasks can AI fully or partially handle? Mark as Full, Partial, or None* | *What should the human do with the freed-up time? Define the new high-value activities* | *What skills does the new role require that the current role doesn't?* | *How will we close the gap? Courses, coaching, on-the-job learning, hiring?* |
| **Example: Relationship Manager**<br>Client data gathering (25%)<br>Report preparation (20%)<br>Compliance docs (15%)<br>Client meetings (25%)<br>Prospecting (15%) | Full: AI pulls from systems<br>Full: AI generates drafts<br>Partial: AI pre-fills<br>None, human essential<br>Partial: AI identifies leads | Deeper client discovery conversations<br>Complex financial planning<br>Proactive outreach on life events<br>Coaching junior advisors<br>Building referral networks | Advanced consultative skills<br>Behavioral finance basics<br>AI output interpretation<br>Empathetic communication<br>Coaching and mentoring | 2-day consultative skills workshop<br>Monthly AI tools training<br>Peer coaching program<br>Quarterly role reviews with manager |

*Figure 3.1: The Role Redesign Canvas. Example for Relationship Managers*

The canvas forces several things that casual thinking about AI and roles tends to skip. First, it makes the time allocation visible. When you see that 60 percent of a relationship manager's time goes to tasks AI can handle, the scale of the transformation becomes real. Second, it makes the new role explicit. "More time for high-value activities" is vague; "deeper client discovery conversations, proactive life-event outreach, and coaching junior advisors" is actionable. Third, it connects role change to skills and training, so the support isn't an afterthought.

I recommend doing this exercise collaboratively, with the people who actually do the work, not just their managers. Frontline staff know their tasks better than anyone, and involving them in the redesign process is itself a powerful change management tool. People who help design their own future are far more likely to embrace it than people who have a future imposed on them.

• • •

## The Elephant in the Room: Automation Anxiety

Let's talk about the thing everyone in the organization is thinking about but few are discussing openly: fear.

Fear of job displacement is consistently the single biggest barrier to AI adoption among employees in financial services. And it's not irrational. When people see AI handling tasks they used to do, the natural question is: "If the machine can do my job, why do they need me?" No amount of corporate messaging about "augmentation" and "upskilling" will eliminate that fear if employees don't see concrete evidence that the institution is invested in their future.

The institutions that handle this well share three characteristics.

### Radical honesty about what's changing

People can handle hard truths. What they can't handle is the feeling that they're being lied to. When leadership says "AI won't affect any jobs" while simultaneously deploying systems that clearly automate significant portions of those jobs, employees don't believe the reassurance. They just stop trusting leadership.

Better to be honest: "Some of what you do today will be done by AI in the coming months. Here's what that means. Here's what your role will look like. Here's how we're going to support you through the transition. And here's what we need from you." Honesty doesn't eliminate anxiety, but it creates the conditions for trust. And trust is the only foundation on which successful change management can be built.

## Visible and credible upskilling paths

It's not enough to say "we'll retrain you." People need to see the path. What specific skills will they learn? What does the new role look like in practice? Who has already made the transition, and what was their experience? Are there certifications, career progressions, and compensation adjustments that reflect the new role's value?

The most effective programs I've seen combine formal training (workshops, courses, certifications) with experiential learning (paired work with AI tools, gradually expanding scope) and visible role models (peers who have made the transition and can speak to it credibly). When an operations analyst can point to a colleague who went from manual processing to exception analysis and is now earning more and doing more interesting work, the abstract promise of "upskilling" becomes concrete and believable.

## The "no surprises" principle

Involve employees in AI design and deployment, not just as end-users but as co-designers. When contact center agents help define how the AI virtual assistant should escalate calls to humans, they have ownership of the system and insight into how it works. When operations staff participate in training the AI on edge cases, they see firsthand that the AI needs them, that their expertise is being built into the system, not replaced by it.

The "no surprises" principle also means that teams learn about AI changes before they experience them. Not in a company-wide email that nobody reads, but in team conversations, with their direct manager, with the opportunity to ask questions and voice concerns. The investment in this communication seems expensive when you're moving fast. The cost of not making it (in resistance, turnover, and failed adoption) is far higher.

• • •

## New Leadership Expectations

If AI changes what frontline roles look like, it also changes what leadership looks like.

In a pre-AI environment, a good manager in financial services was someone who could drive performance against clear metrics, enforce process discipline, and develop people along well-understood career paths. Those skills still matter. But AI adds new dimensions that many current leaders weren't trained for and don't feel confident navigating.

Three leadership capabilities stand out as especially important.

### Coaching people to work with AI

This is more than "showing people how to use the new system." It's about helping people develop judgment about when to trust AI outputs and when to question them. It's about normalizing the experience of uncertainty, because working with AI inherently involves situations where the right answer isn't obvious and the technology doesn't have all the context. And it's about creating a team culture where people feel comfortable saying "I don't trust this AI output" without being seen as Luddites or obstructionists.

### Managing hybrid human-AI teams

When AI is a teammate, what does team management look like? Leaders need to think about how work is allocated between humans and AI. They need new performance metrics that reflect the new reality, measuring the quality of human-AI collaboration, not just human output. They need feedback loops that capture how well the AI is performing and how effectively humans are overseeing it. This is unfamiliar territory for most managers, and they need support (frameworks, training, and peer learning) to navigate it.

### Modeling the right behaviors

Employees take their cues from their leaders. If a leader is visibly curious about AI, willing to experiment, honest about what they don't know, and open about mistakes, their team will mirror those behaviors. If a leader treats AI as someone else's problem, avoids using the tools, or dismisses concerns as resistance, their team will disengage.

The most effective AI leaders I've observed share a combination of humility ("I'm learning this too"), curiosity ("Let's figure this out together"), and ethical grounding ("We're going to do this in a way that's fair and transparent"). None of these require technical expertise. All of them require deliberate effort.

• • •

## Change Readiness Checklist

Before deploying AI into a team's workflow, I recommend leaders work through the following ten questions. They're not a pass-fail test; they're a diagnostic. A "no" answer doesn't mean you can't proceed; it means you've identified a gap that needs attention.

| # | Question | Y / N / Partial |
|---|---|---|
| 1 | Have we clearly defined how AI will change this role's tasks and responsibilities? | |
| 2 | Do team members understand what will change and what won't? | |
| 3 | Have we redesigned the role (not just removed tasks), including new high-value activities? | |
| 4 | Do updated performance metrics reflect the new role, not just the old one? | |
| 5 | Is there a concrete upskilling plan with timelines, resources, and visible career paths? | |
| 6 | Have team members been involved in the design or piloting of the AI tools? | |

| # | Question | Y / N / Partial |
|---|---|---|
| 7 | Can team members articulate what the AI does and what its limitations are? | |
| 8 | Does the team leader feel confident coaching their team through the transition? | |
| 9 | Is there a feedback mechanism for employees to flag concerns or suggest improvements? | |
| 10 | Have we communicated honestly about the change, including what we don't yet know? | |

*Figure 3.2: Change Readiness Checklist. Ten Questions Before Deploying AI into a Team*

If you can answer "yes" to seven or more of these questions, your team is likely ready for a well-managed AI deployment. Five or six, and you should proceed only with targeted preparation in the gaps you've identified. Fewer than five, and you're deploying technology into a team that isn't prepared to absorb it, which means you're likely to see resistance, workarounds, and disappointing adoption, no matter how good the technology is.

The checklist also serves a secondary purpose: it gives leaders a structured way to have the conversation with their teams. Walking through these questions together (openly, honestly) is itself an act of people-centric leadership. It signals that the organization cares about the human side of the change, not just the technical side.

• • •

## The Work After the Work

I want to return to Sarah for a moment.

Six months after her bank's initial AI deployment, things had improved, but only because her branch manager took the role redesign seriously. He sat down with each relationship manager individually. They mapped their current tasks, identified what AI had changed, and defined, together, what the new role would look like. He advocated for updated performance metrics. He arranged for the team to attend a consultative skills workshop. He created a weekly "AI check-in" where the team shared what was working, what wasn't, and what they were learning.

Sarah's anxiety didn't disappear overnight. But it was replaced, gradually, by something more productive: a sense that she was evolving into a more valuable version of herself. That the skills she was developing (deeper client engagement, complex problem-solving, ethical judgment) were exactly the skills that would matter most as AI continued to advance. That she wasn't being replaced. She was being upgraded. That is what it means to keep the human in the lead, not resisting AI, but shaping it so that people grow alongside it.

That outcome wasn't inevitable. It was designed. And it required a leader who understood that deploying AI into a team without redesigning the work, supporting the people, and managing the change isn't an AI initiative. It's a disruption disguised as innovation.

In the next chapter, we shift our focus from employees to the other group most directly affected by AI in financial services: customers. How do we design AI-powered experiences that deepen trust and serve customers better, including the ones who are most vulnerable? That's where we're headed.

CHAPTER 4

# Designing Customer-First AI Journeys

Two customers. Same bank. Same week. Radically different experiences.

The first is Elena, a thirty-eight-year-old marketing director who recently had her second child. She logs into her banking app on a Tuesday morning and finds a notification waiting: "On maternity leave? Here are three things to review about your mortgage protection, life insurance coverage, and emergency savings." The suggestions are specific, relevant, and timely. The links take her to clear explanations, not sales pages. There's an option to schedule a call with an advisor if she wants to talk it through. Elena doesn't schedule the call (she doesn't need to), but she screenshots the notification and texts it to her partner. "Our bank actually gets it," she writes.

The second is Marcus, a sixty-three-year-old small business owner who's been a customer of the same bank for over twenty years. He applied for a credit line increase to manage seasonal cash flow, something he's done before, always successfully. This time, his application was declined within minutes. The message said: "Based on our assessment, we're unable to approve your request at this time." No explanation. No detail. No mention of what "assessment" meant or what factors were considered. When Marcus called the branch, his relationship manager couldn't explain the decision either. "The system flagged it," she said, clearly uncomfortable. "I'm not sure why."

Marcus wasn't just frustrated. He was humiliated. He'd built his business over decades and maintained an impeccable banking relationship; now a machine had judged him unworthy, and nobody could tell him why. He started looking for a new bank that week.

These two stories capture the spectrum of what AI-powered customer experience looks like in financial services today. When it works, it feels almost magical, as if the institution understands you as a person, not just an account number. When it fails, it's worse than having no AI at all, because the opacity of automated decisions amplifies the sense of powerlessness.

This chapter is about designing for Elena's experience and preventing Marcus's. It's about the principles, frameworks, and hard design choices that determine whether AI makes customers feel seen or surveilled, supported or abandoned.

• • •

## Hyper-Personalization with Guardrails

The promise of AI-driven personalization in financial services is extraordinary. Instead of treating every customer as a segment ("millennials," "high-net-worth," "small business"), AI can tailor advice, offers, and experiences to individual needs, behaviors, and life stages. The technology is already there. McKinsey's Next in Personalization research shows that 71 percent of consumers expect personalized interactions, and 76 percent get frustrated when they don't receive them. McKinsey's research also finds that companies that grow faster drive 40 percent more of their revenue from personalization than their slower-growing peers.

But personalization in financial services isn't like personalization in retail. When Netflix recommends a movie you don't like, you lose ninety minutes. When your bank recommends a financial product that doesn't fit your situation, or worse, when an algorithm makes a consequential decision about your creditworthiness based on data you didn't know it was using, the stakes are fundamentally different.

That's why effective personalization in financial services requires what I call a consent-and-clarity contract. It has three elements.

## Transparency about data use

Customers should understand what data is being used to personalize their experience. Not in a thirty-page privacy policy that nobody reads, but in plain language at the point of interaction. Elena's notification about maternity-related financial planning was powerful because it felt relevant. But if Elena had no idea her bank was using her spending patterns, account activity, and possibly third-party data to infer that she'd recently had a child, that same notification could feel invasive. The difference between helpful and creepy is transparency.

The best implementations I've seen provide a brief, human-readable explanation alongside personalized recommendations. Something like: "We noticed changes in your account activity that often accompany a growing family. Here are some things to consider." It's honest. It acknowledges the inference. And it gives the customer the context they need to evaluate the recommendation.

## Meaningful opt-out mechanisms

Consent isn't a one-time checkbox during account opening. It's an ongoing relationship. Customers should be able to see what personalization is active, what data powers it, and how to turn it off, partially or completely, without losing access to core banking services. This is both an ethical imperative and, increasingly, a regulatory one. Europe's privacy and AI rules (GDPR's consent requirements and the AI Act's transparency obligations), together with emerging frameworks worldwide, are pushing financial institutions toward granular, revocable consent for AI-driven personalization.

### Fairness testing as a continuous discipline

Here's the uncomfortable truth about personalization: the same algorithms that tailor experiences to individual needs can also discriminate. If the model that generates Elena's helpful notification is systematically less helpful to customers of certain demographics, offering fewer proactive suggestions, less favorable product recommendations, or fewer opportunities to engage with human advisors, then personalization has become discrimination, regardless of intent.

Fairness testing means regularly auditing AI-driven personalization for differential treatment across protected characteristics. It means comparing outcomes, not just inputs, across customer segments. And it means having the institutional courage to pull back a feature that's generating strong business results if it's doing so in a way that disadvantages certain groups. We'll explore this in greater depth in Chapter 9, where we watch hidden bias surface in a fraud detection system, but the principle belongs here too: personalization that isn't fair isn't customer-first. It's just sophisticated segmentation with new risks attached.

• • •

## Conversational Banking Done Right

The rise of AI-powered virtual assistants (chatbots, voice assistants, and, increasingly, agentic AI that can take actions on behalf of customers) represents one of the most visible changes in how banks interact with their customers. The numbers are significant: AI chatbots now handle a substantial share of initial customer interactions at major banks, and by one market tracker's estimate, the global chatbot market has surpassed eleven billion dollars, nearly doubling in five years.

But the track record is, at best, mixed. The US Consumer Financial Protection Bureau has flagged serious concerns about chatbot performance in banking, documenting cases where customers received inaccurate information, got trapped in circular conversation loops, or were unable to reach a human agent when they needed one. In late 2025, a high-profile investigation by the UK consumer group Which? found that major AI chatbots were giving consumers actively dangerous financial advice, recommending they exceed legal contribution limits on tax-advantaged accounts, providing incorrect tax guidance, and directing them to paid services when free government alternatives existed.

And in early 2026, a security researcher at TELUS Digital published tests of twenty-four AI models configured as banking customer-service assistants and found that every single one was exploitable, with manipulation success rates ranging from 1 percent to over 64 percent.

These aren't edge cases. They're predictable failures that occur when general-purpose AI is deployed in a high-stakes, heavily regulated domain without sufficient guardrails, testing, and human backup. The question isn't whether to use conversational AI in banking (the efficiency gains are too significant to ignore), but how to do it in a way that protects customers and builds trust.

Three design principles make the difference.

## Design for the moments that matter

Not all customer interactions are created equal. Checking a balance is low-stakes and routine. Reporting a stolen card is high-stakes and emotional. Asking whether to consolidate debts is complex and consequential. The best conversational banking implementations recognize these distinctions and design for them deliberately.

For routine queries (balances, transaction history, branch hours, payment status), AI virtual assistants work beautifully. They're fast, accurate, available around the clock, and most customers prefer them. It's for the other moments (the stressful, confusing, or high-consequence ones) that the design has to be different. These are the moments when trust is won or lost, and they require what I call escalation intelligence: the ability for the AI to recognize its own limits and transition seamlessly to a human.

## The escalation design principle

Make the path to a human easy to find, never buried, when the conversation exceeds AI capability or when emotions are high. This sounds simple, but it runs directly counter to the cost optimization logic that drives many chatbot deployments. If the goal is to reduce the cost per customer interaction, then every conversation that escalates to a human represents a failure. This incentive structure produces the worst possible design: chatbots that work hard to keep you talking to a robot, even when it's clearly not working.

The customer-first approach inverts this logic. Escalation to a human isn't a failure; it's a feature. It's the system working as designed. The AI handles what it's good at and hands off what it's not, seamlessly, with full context transferred so the customer doesn't have to repeat themselves. The best implementations I've observed use sentiment detection, topic complexity scoring, and explicit customer signals to trigger escalation proactively, before the customer has to ask, and certainly before they become frustrated enough to escalate on their own.

## Guardrails for regulated content

Financial services conversations frequently involve topics where inaccurate information can cause real harm: tax implications, regulatory limits, product suitability, insurance coverage. A virtual assistant that confidently provides wrong information about contribution limits for ISAs (the United Kingdom's tax-advantaged savings accounts) or mischaracterizes the terms of a financial product isn't just annoying; it's potentially illegal and almost certainly harmful.

The guardrail approach means defining clear boundaries for what the AI can and cannot advise on, testing rigorously within those boundaries, and designing explicit fallback behavior for topics that require human expertise or regulatory precision. This requires ongoing investment; it is not something you configure once and forget, because the regulatory landscape evolves, products change, and AI models can drift in unpredictable ways.

• • •

## Balancing Digital Self-Service with Human Access

There's a spectrum of customer interaction in modern banking, from fully automated to fully human, with several stages in between. The strategic question isn't which end of the spectrum to be on; it's how to let customers move along it based on their needs, preferences, and circumstances.

Roughly half of customers, depending on the survey, say they're willing to use AI-powered assistants for service interactions. That's a significant number, but it also means roughly half prefer alternatives. Customer-first AI design respects this diversity rather than forcing everyone through the same channel.

The most thoughtful implementations I've seen offer what I call a channel spectrum with visible exit ramps. The default might be a capable AI assistant, but at every point in the interaction, the customer can see, and easily access, the option to speak with a human. There's no hunting through menus, no "that option is not available right now," no subtle friction designed to keep you in the automated channel.

This is especially critical for one group that deserves particular attention: vulnerable customers.

### Designing for vulnerability

In financial services, vulnerability takes many forms. There are elderly customers who may not be comfortable with digital interfaces. There are people experiencing financial distress (job loss, debt crisis, the financial fallout of illness or bereavement) who need empathy and flexibility that AI cannot reliably provide. There are customers with cognitive disabilities, language barriers, or limited digital literacy. And there are people who are simply having the worst day of their lives and need a human being to talk to.

For these customers, AI can be either a bridge or a barrier. A bridge when it's designed to detect vulnerability signals (unusual transaction patterns, repeated failed interactions, specific language cues) and route those customers to specialized human support. A barrier when it creates an obstacle course that a distressed or confused customer can't navigate.

The UK's Financial Conduct Authority (FCA) has been particularly clear on this point: firms must ensure that their use of AI doesn't create new barriers for vulnerable customers or remove the human interactions that vulnerable customers rely on. This isn't just regulatory compliance; it's a moral imperative. The measure of a financial institution's character isn't how it treats its most profitable, tech-savvy customers. It's how it treats the ones who are struggling.

## Accessibility and inclusion as design requirements

Accessibility is not an afterthought or a compliance checkbox. It's a design requirement that should be baked into AI-powered customer experiences from the start. This means screen reader compatibility for visual interfaces. Alternative interaction modes for customers who can't use chatbots or voice assistants. Clear, jargon-free language. Support for multiple languages. Font sizes and contrast ratios that work for aging eyes. Transaction confirmations that are comprehensible, not just legally sufficient.

It also means testing with real users from diverse backgrounds and ability levels, not just the thirty-five-year-old, digitally native persona that too often stands in for "the customer" in product design. The digital divide in banking is real, and AI has the potential to either narrow it or widen it, depending entirely on design choices.

• • •

## Measuring What Matters for Customers

Here's a pattern I've seen at institution after institution: the team deploys an AI chatbot or personalization engine, and the primary success metrics are call-time reduction, cost per interaction, and deflection rate, the percentage of inquiries handled without a human. These are efficiency metrics, and they're important. But they tell you almost nothing about whether customers are actually being well served.

A customer who hangs up in frustration counts as a "resolved" interaction in some measurement systems. Consider a customer who wanted to speak to a human but was deflected to a chatbot that couldn't help. That counts as a successful deflection. A customer who received personalized recommendations but felt their privacy was violated, well, the click-through rate looked great.

Customer-first AI requires customer-first measurement. That means supplementing efficiency metrics with three additional categories.

### Trust metrics

Do customers feel their bank is acting in their interest? Trust is the foundational currency of financial services, and AI can either build it or erode it. Measuring trust requires going beyond satisfaction surveys to understand how customers feel about specific AI-powered interactions. Did the personalized recommendation feel helpful or intrusive? Was the chatbot interaction efficient or frustrating? When the algorithm made a decision, did the customer feel they understood why? Salesforce's *State of the Connected Customer* survey found as far back as 2019 that 63 percent of customers believe most companies fail to use their data transparently. Institutions that don't measure and respond to this gap are flying blind.

### Resolution quality

Not just whether the interaction was resolved, but how well. Did the customer get accurate information? Was the issue fully addressed, or did they need to call back? Did the resolution happen at a pace that respected the customer's time? And critically: did the AI-to-human handoff (when it happened) work smoothly, or did the customer have to start over? Resolution quality is harder to measure than deflection rate, but it's far more predictive of customer loyalty and long-term value.

### Vulnerable-customer outcomes

Are AI systems helping or hindering those who need the most support? This means tracking outcomes specifically for customer segments identified as potentially vulnerable: elderly customers, those in financial difficulty, and those with accessibility needs. Are these customers getting through to human support when they need it? Are they receiving the same quality of proactive outreach? Are there systematic differences in how AI-driven decisions affect them compared to the general population?

If you're only measuring efficiency, you can accidentally optimize for a system that works brilliantly for your easiest customers and fails your most vulnerable ones. Customer-first measurement catches this.

• • •

## The Customer AI Journey Map

When I work with teams on customer-facing AI, one of the most useful exercises is mapping the customer journey and classifying each touchpoint according to the role AI should play. The template below provides the structure.

| Touchpoint | Customer Need | Ideal AI Role | Human Role | Escalation Trigger | Success Metric |
|---|---|---|---|---|---|
| **Account opening** | Quick, clear, guided | Pre-fill forms, verify identity, explain options | Complex eligibility, product suitability review | Eligibility questions, multiple attempts | Completion rate, time to open, abandonment rate |
| **Routine service inquiry** | Fast, accurate answer | Fully automated: balance, transactions, FAQs | None needed for standard queries | 2+ failed attempts, frustration signals | First-contact resolution, accuracy |
| **Financial advice / planning** | Trusted, personalized guidance | Prepare brief, surface options, draft plan | Lead the conversation, exercise judgment, build trust | Always human-led; AI supports | Client satisfaction, advice quality, relationship depth |
| **Complaint or dispute** | Feeling heard, fair resolution | Log details, retrieve history, suggest resolution | Empathy, judgment, authority to resolve | Immediately; complaints always routed to human | Resolution time, customer retention, fairness |
| **Fraud report or financial distress** | Urgent help, reassurance, protection | Detect, alert, freeze; gather initial info | Specialized agent with authority and empathy training | Immediately; vulnerability detected | Response time, resolution, customer well-being |
| **Proactive outreach** | Relevant, timely, respectful | Identify life events, trigger notifications, personalize | Available for follow-up conversation | Customer requests human; complex life event | Engagement rate, trust score, opt-out rate |

*Figure 4.1: Customer AI Journey Map Template. Mapping AI and Human Roles Across Touchpoints*

The power of this exercise is in the conversations it forces. When you map every touchpoint, you start to see patterns. You notice that the highest-stakes moments, the ones that most determine whether a customer stays or leaves, are almost always the ones where human involvement is essential. You see that AI's greatest value isn't in replacing human interactions but in making them better: giving the human agent more context, more preparation time, more insight into what the customer actually needs.

You also see the escalation triggers clearly. And you can design for them proactively rather than discovering them reactively, through customer complaints, social media posts, or regulatory inquiries.

• • •

## The Escalation Design Checklist

Escalation from AI to human is the most critical design element in customer-facing AI. Get it right, and customers feel supported. Get it wrong, and they feel trapped. These eight principles guide effective escalation design.

| # | Principle | What It Means in Practice |
|---|---|---|
| 1 | Visibility | The option to reach a human is always visible in the interface, never hidden behind three menus or available only after the chatbot has failed. |
| 2 | No dead ends | The AI never says "I can't help with that" without immediately offering a human alternative. Every conversation has a path forward. |
| 3 | Context transfer | When a customer is transferred from AI to human, the full conversation history goes with them. The customer never has to repeat themselves. |
| 4 | Proactive detection | The system uses sentiment analysis, complexity scoring, and behavioral signals to identify when escalation is needed, before the customer asks. |
| 5 | Emotional sensitivity | Conversations involving distress, confusion, or frustration are escalated immediately. The AI does not attempt to "fix" emotional situations. |

| # | Principle | What It Means in Practice |
|---|---|---|
| 6 | Specialist routing | Escalation goes to the right human, not just any available agent, but one with relevant expertise (fraud specialist, financial advisor, accessibility support). |
| 7 | Speed guarantee | Escalated customers are prioritized, not placed at the back of the queue. The wait time is communicated honestly and updated in real time. |
| 8 | Feedback loop | Every escalation is logged and analyzed. Patterns of escalation drive improvements to the AI, reducing the need for future escalations organically. |

*Figure 4.2: Escalation Design Checklist. Eight Principles for Effective Human Escalation*

The checklist seems straightforward, but most implementations violate at least three of these principles. The most common failure? Number one, hiding the exit. When cost reduction is the primary KPI, there's an institutional incentive to make the chatbot sticky, even when stickiness means frustration. Customer-first design resists that incentive.

I recommend teams use this checklist during the design phase, not after launch. Walk through it with real customer scenarios, especially the difficult ones, the edge cases, the moments of distress. If the design doesn't pass all eight principles for those scenarios, it isn't ready.

• • •

## The Trust Equation

Everything in this chapter comes back to a single idea: trust.

Trust is the foundational asset of financial services. It takes years to build and moments to destroy. And AI, for all its power, is a trust amplifier; it makes existing trust dynamics stronger, in whichever direction they're already heading. An institution that customers already trust can use AI to deepen that trust, delivering experiences that feel personal, timely, and genuinely helpful.

An institution where trust is fragile will find that AI's opacity, speed, and scale amplify customers' unease.

Elena trusts her bank more after her AI-powered notification because the experience confirmed what she already believed: that her bank understands her and has her interests at heart. Marcus trusts his bank less after his algorithmic credit denial because the experience confirmed his growing suspicion that he's become a number in a system that doesn't know him and doesn't care. Within a month, he moved his twenty-year relationship to a competitor that could explain its decisions and still answered the phone with a person.

Same technology. Same bank. Opposite outcomes. The difference is design, design that keeps the human in the lead, not just behind the dashboard but at the center of every customer experience.

Customer-first AI design isn't a luxury. It isn't a nice-to-have that you layer on after the core technology is working. It's the difference between AI that strengthens your customer relationships and AI that quietly unravels them. And in a world where a large majority of customers say they'll drop a brand over a single unresolved issue, the margin for error is vanishingly thin.

In the next chapter, we step behind the curtain to look at the governance structures, controls, and oversight mechanisms that make trustworthy customer-facing AI possible. Because delivering the experience Elena had, and preventing the experience Marcus had, doesn't happen by accident. It requires robust governance foundations. That's where we're headed next.

CHAPTER 5

# Governance Foundations

## *Controls, Oversight, and Accountability*

Imagine a regulatory examiner walks into your bank. She's not hostile, just thorough. She's here to review your AI lending model, the one that processes 60 percent of your consumer credit decisions. She sits down with your chief risk officer and asks three questions.

> *"Who approved this model for deployment?"*
>
> *"How do you know it's treating customers fairly?"*
>
> *"What happens when it goes wrong?"*

At one bank, the CRO can answer each question without hesitation. She points to a formal approval by the model risk committee, names the independent validation team, and walks through the fairness testing methodology. She pulls up the AI system inventory, a live record that shows every AI model in production, its risk classification, its owner, and its last review date. She explains the incident management playbook: predefined escalation paths, rollback procedures, and communication protocols. The examiner nods, takes notes, and moves on.

At another bank, the answers come more slowly. The model was approved by... the data science team? Maybe with a sign-off from the business unit head. Fairness testing was done at launch, the CRO thinks, but he's not sure whether it's been repeated since. When the examiner asks about incident management, there's a pause. "We haven't had an incident yet," he says, as if that were reassuring. The examiner makes a different kind of note.

Both banks deployed the same type of AI technology. Both achieved similar efficiency gains. But only one had built the governance foundations to sustain it. The other was operating on borrowed time.

This chapter is about building those foundations. Not as a bureaucratic exercise or a compliance checkbox, but as the infrastructure that makes responsible AI possible at scale. Because in financial services, governance isn't the thing that slows innovation down. It's the thing that makes innovation trustworthy enough to last.

• • •

## Governance Starts at the Top

Let me be direct about something that too many organizations get wrong: AI governance is a board and executive leadership responsibility. It is not a technology team deliverable.

This isn't just my opinion. It's the emerging consensus among regulators worldwide. The European Central Bank, after conducting workshops with thirteen major banks in 2025, observed that while about half of banks in its sample had introduced dedicated AI policies or committees, far more work was needed to ensure that second and third lines of defense, risk management and internal audit, were adequately overseeing AI use. Some banks had appointed dedicated senior roles like a Chief AI Officer, but the message was clear: governance of AI needs to be embedded into existing leadership structures, not bolted on beside them.

The EU AI Act reinforces this point. Under its provisions, high-risk AI systems in the financial sector must comply with specific requirements around risk management, data governance, human oversight, and documentation. These aren't requirements that a data science team can fulfill alone. They require decisions about risk appetite, resource allocation, organizational accountability, and strategic priorities, decisions that belong to the board and the executive committee.

What does board-level AI governance actually look like in practice? Three things.

## Strategic clarity on AI risk appetite

The board needs to articulate what level of AI risk the institution is willing to accept, in which areas, and under what conditions. This doesn't mean the board needs to understand the mathematics of gradient boosting. It means they need to make informed decisions about questions like: Are we comfortable using AI for credit decisions? For fraud detection? For customer-facing advice? What's our tolerance for model error? What's our exposure if a model behaves in an unexpected way? These are fundamentally strategic questions, and they require the same rigor and deliberation as any other major risk decision.

## Accountability structures that don't depend on goodwill

Every AI system in production needs a named owner, someone who is accountable for its performance, its compliance, and its outcomes. Not the team that built it. Not the vendor that sold it. A specific individual with the authority and the incentive to ensure the system operates as intended. In too many organizations, AI accountability is diffuse to the point of being nonexistent. The data science team built the model. The business unit deployed it. IT maintains the infrastructure. Compliance reviewed it, once. When something goes wrong, everyone points to everyone else. Clear accountability structures prevent this.

## Board-level AI literacy

Directors don't need to become data scientists. But they need enough understanding to ask the right questions, challenge management's assertions, and recognize when they're being given incomplete or misleading information. What kind of AI are we deploying? What data does it use? How was it validated? What could go wrong? How would we know? These aren't technical questions; they're governance questions. And a board that can't ask them is a board that can't govern AI effectively.

• • •

## Core Control Themes

Once the governance structure is in place at the top, the practical work begins: building the control environment that makes AI trustworthy in operation. Five themes form the core of that environment.

### Risk assessment and classification

Not all AI systems carry the same risk. A model that recommends which marketing email to send is not the same as a model that decides whether someone gets a mortgage. Proportionate governance, applying heavier controls to higher-risk systems and lighter controls to lower-risk ones, is both efficient and effective. It's also the approach enshrined in the EU AI Act, which classifies AI systems into risk tiers with corresponding obligations.

The practical challenge is building a classification system that's rigorous enough to be meaningful but simple enough to actually use. Later in this chapter, I'll share an AI System Classification Template that provides the structure. But the principle is straightforward: assess each AI system against a set of risk factors (impact on customers, regulatory sensitivity, autonomy of decisions, reversibility of outcomes) and assign proportionate governance accordingly.

### Data quality governance

AI is only as good as its data. This statement has become a cliché, but it's a cliché because it's relentlessly true. A credit scoring model trained on biased historical data will produce biased outcomes. A fraud detection system fed incomplete or stale data will miss patterns or generate false positives. A customer personalization engine using inaccurate data will make recommendations that feel wrong or invasive.

Data quality governance for AI means ensuring the integrity, completeness, and appropriateness of data at three stages: the training data used to build the model, the ongoing data feeds that keep it current, and the data lineage that documents where data came from, how it was transformed, and what assumptions were made. The mapping exercise by the European Banking Authority (EBA) confirmed that existing data governance frameworks in banking, particularly those under the Capital Requirements Regulation, provide a solid foundation. The task isn't to build something new from scratch. It's to extend what already exists to cover the specific demands of AI.

## Model validation

Independent validation of AI models before deployment is not optional. It's the most fundamental control in the AI governance toolkit. Validation means testing the model's performance, accuracy, and fairness using data it hasn't seen. It means stress-testing it against edge cases and adversarial scenarios. It means having someone other than the model's creators assess whether it's fit for purpose. And it means documenting everything: the methodology, the results, the limitations, and the conditions under which the model should be re-validated.

For institutions with established model risk management frameworks, this is familiar territory. The challenge with AI and machine learning models is that they can be less interpretable than traditional statistical models, which makes validation both more important and more difficult. We'll address this tension in Chapter 8 when we discuss explainability, but the governance principle is simple: no AI model goes into production without independent validation. Period.

## Monitoring and drift detection

Validation at deployment is necessary but not sufficient. AI models degrade over time. The world changes (customer behaviors shift, economic conditions evolve, regulatory requirements update), and a model that performed well at launch may not perform well six months later. This is model drift, and it's one of the most underappreciated risks in AI deployment.

Effective monitoring means tracking model performance continuously against defined thresholds. It means watching for data drift, changes in the statistical properties of input data that may signal the model is operating outside its training distribution. And it means having predefined triggers for re-validation: when performance drops below a threshold, when the operating environment changes significantly, or at regular intervals regardless. The institutions that get this right treat monitoring as a core operational function, not an afterthought.

## Incident management

Every AI system will eventually produce an unexpected, incorrect, or harmful output. The question isn't whether it will happen, but whether you're ready when it does. Incident management for AI means having predefined playbooks that specify: who gets notified, what decision rights they have, how quickly the system can be rolled back or overridden, how affected customers are identified and remediated, and how the incident is documented and analyzed for root cause.

The best incident management frameworks I've seen draw on operational resilience practices that financial institutions already have, and for good reason. The EBA's mapping exercise found that the requirements of the Digital Operational Resilience Act (DORA) for cybersecurity and business continuity extensively cover the AI Act's relevant provisions. Once again, the message is: don't reinvent the wheel. Extend what you have.

• • •

## Aligning AI Governance with Existing Frameworks

One of the most common, and most expensive, mistakes in AI governance is creating an entirely parallel governance structure. A new committee. A new policy framework. A new set of reporting lines. A new audit program. All dedicated to AI, all separate from the existing risk, compliance, and audit architecture.

This approach fails for three reasons. First, it's inefficient; it duplicates work, creates jurisdictional confusion, and consumes resources that could be better deployed elsewhere. Second, it creates gaps, because AI governance doesn't exist in a vacuum. An AI lending model doesn't just need AI governance; it needs credit risk governance, fair lending compliance, data privacy controls, and operational resilience. Siloing AI governance from these existing frameworks guarantees incomplete coverage. Third, it breeds resistance; adding another governance layer on top of what already exists is how you ensure that the people doing the work see governance as an obstacle rather than a support.

The better approach, and the one that regulators are increasingly endorsing, is integration. The EBA's landmark finding in its AI Act mapping exercise was unambiguous: no significant contradictions exist between the AI Act and EU banking regulation. The AI Act is complementary to existing banking legislation, which already provides a comprehensive framework for managing risks. The Capital Requirements Regulation and Directive provide a technology-neutral governance and risk management framework that can be applied to AI. DORA covers cybersecurity and operational resilience. Existing model risk management frameworks cover validation and monitoring.

What this means in practice is that AI governance should be an extension of, not a replacement for, your existing governance architecture. Three integration points are particularly important.

## Extending model risk management to AI/ML

Most large financial institutions already have model risk management frameworks that govern traditional statistical models. Extending these to cover machine learning models is the most natural and effective starting point. The core principles are the same: independent validation, ongoing monitoring, documentation, and governance. What changes is the technical methodology. ML models may require different validation techniques, more frequent monitoring, and more attention to explainability. But the governance structure doesn't need to be reinvented.

## Integrating with operational resilience

AI systems are operational systems. They need the same resilience controls as any other critical technology: availability, recoverability, incident management, and third-party risk management. For institutions that have already invested in DORA compliance, much of this infrastructure is in place. The additional AI-specific considerations (model rollback capabilities, fallback decision processes, and AI-specific vendor oversight) can be layered onto existing resilience frameworks.

## Mapping AI governance to regulatory expectations

The regulatory landscape for AI in financial services is converging rapidly. The EU AI Act, the EBA's supervisory approach, the ECB's priorities for 2026–2028, and national regulators' guidance are all coalescing around a consistent set of expectations: risk-based classification, documentation and transparency, human oversight, fairness and non-discrimination, and robust governance structures. Institutions that map their AI governance to these expectations, rather than treating each regulatory requirement as a separate compliance exercise, will find the burden manageable and the result coherent.

• • •

## Making AI "Audit-Ready"

If I could give one piece of practical advice to every institution deploying AI, it would be this: build documentation as you go, not after the fact.

Documentation is the currency of accountability. Without it, governance is aspirational rather than real. When a regulator asks "how do you know this model is fair?" the right answer isn't "our data scientists told us." It's a documented record of the fairness testing methodology, the results, the review and approval process, and the ongoing monitoring plan. When an auditor asks "what AI systems do you have in production?" the right answer isn't "let me check with the technology team." It's a comprehensive, up-to-date AI system inventory.

Four documentation practices are essential.

### The AI system inventory

You cannot govern what you cannot see. Yet a surprising number of institutions cannot produce a complete list of the AI systems they have in production, where they're deployed, who's accountable for them, and when they were last reviewed. The AI system inventory is the starting point for everything else. It should include: the system's purpose and risk classification, its data inputs and outputs, its owner and oversight committee, its validation and review history, and its current performance against defined metrics.

### Decision documentation

At every stage of an AI system's lifecycle (design, development, validation, deployment, monitoring, retirement), decisions are made. Why was this algorithm chosen over alternatives? What trade-offs were accepted? What assumptions were made about the data? What limitations were identified? Who approved deployment, and on what basis? These decisions need to be documented, not because someone will be blamed, but because the rationale matters. When conditions change or outcomes surprise, you need to understand the decisions that led to the current state.

### Decision logs for AI outputs

This is a subtle but important distinction from decision documentation about the AI system itself. Decision logs capture what the AI decided in specific cases, why (to the extent that's traceable), and what a human reviewer thought about it. In a lending context, this means logging not just the AI's recommendation (approve/decline) and the factors that drove it, but also the human reviewer's assessment, whether they agreed, disagreed, or overrode the model, and why. These logs are invaluable for detecting patterns of disagreement between human and AI judgment, for fairness audits, and for demonstrating to regulators that human oversight is real, not ceremonial.

### Version control and change management

AI models change. They're retrained, updated, fine-tuned, and replaced. Each change carries risk: a retrained model might perform differently from its predecessor in ways that aren't immediately obvious. Rigorous version control means maintaining a complete record of every model version, what changed, why, and what validation was performed before the new version went into production. Change management means applying the same governance rigor to model updates as to initial deployments, proportionate to the risk, but never skipped.

• • •

## AI Governance Maturity Model

Not every institution starts in the same place. Some have sophisticated model risk management frameworks that extend naturally to AI. Others are deploying machine learning models with minimal governance infrastructure. The maturity model below provides a roadmap, not a judgment, for understanding where you are and what it takes to progress.

| Level | Stage | Characteristics | Key Milestones to Advance |
|---|---|---|---|
| 1 | Ad Hoc | AI deployed by individual teams with no central oversight. No inventory of AI systems. Governance is informal or nonexistent. Risk assessment is project-by-project, if it happens at all. | Appoint AI governance owner at executive level. Create initial AI system inventory. Define AI risk appetite statement. Establish basic approval process for new AI deployments. |
| 2 | Emerging | Basic AI policy exists. Some systems are inventoried. Risk classification is in place but inconsistently applied. Validation happens for high-profile models but isn't universal. Documentation is spotty. | Establish AI governance committee with cross-functional representation. Implement consistent risk classification across all AI systems. Require independent validation for all high-risk models. Begin building decision logs. |
| 3 | Defined | Comprehensive AI governance policy integrated with existing risk frameworks. All AI systems inventoried and classified. Independent validation is standard. Monitoring is in place for critical models. Documentation is consistent. | Extend monitoring to all production AI systems. Implement automated drift detection. Build incident management playbooks for AI-specific scenarios. Integrate AI governance with board reporting. |
| 4 | Managed | AI governance is fully integrated with enterprise risk management. Monitoring is continuous and largely automated. Incident management is tested and rehearsed. Board receives regular, meaningful AI risk reporting. Second and third lines actively oversee AI. | Implement predictive model health monitoring. Establish cross-institutional benchmarking. Develop advanced fairness and bias auditing capabilities. Create feedback loops between monitoring insights and model improvement. |

| Level | Stage | Characteristics | Key Milestones to Advance |
| --- | --- | --- | --- |
| **5** | **Optimized** | Governance enables rather than constrains innovation. Risk-based governance accelerates low-risk AI while providing robust oversight for high-risk systems. Continuous learning from incidents and near-misses. Industry thought leadership. | Continuously refine based on evolving regulatory expectations and emerging risks. Share best practices industry-wide. Invest in governance technology and automation. Pursue regulatory sandboxes for innovation. |

*Figure 5.1: AI Governance Maturity Model. Five Levels from Ad Hoc to Optimized*

Most institutions I work with are somewhere between Level 1 and Level 3. If that's where you are, don't be discouraged. The point of the maturity model isn't to make you feel behind; it's to give you a clear picture of what "good" looks like at each stage and what it takes to get to the next one. Progress from Level 1 to Level 3 can happen within twelve to eighteen months with dedicated leadership. Getting from Level 3 to Level 5 is a multi-year journey that requires sustained investment and cultural change.

• • •

## AI System Classification Template

Risk classification is the engine of proportionate governance. The template below provides a practical framework for categorizing AI systems by risk level and determining appropriate governance controls.

| Risk Factor | Low Risk | Medium Risk | High Risk |
| --- | --- | --- | --- |
| **Customer Impact** | No direct customer-facing impact (e.g., internal analytics, operational efficiency) | Influences customer experience but decisions are reversible (e.g., product recommendations, chatbot interactions) | Directly determines customer outcomes with significant consequences (e.g., credit decisions, pricing, fraud blocking) |

| Risk Factor | Low Risk | Medium Risk | High Risk |
|---|---|---|---|
| **Regulatory Sensitivity** | No specific regulatory requirements (e.g., marketing optimization, internal scheduling) | Subject to general regulatory expectations (e.g., data protection, fair treatment) | Subject to specific regulatory requirements (e.g., credit scoring under EU AI Act, AML screening) |
| **Decision Autonomy** | AI provides information only; all decisions made by humans | AI recommends; human approves or overrides (Human-in-the-Loop) | AI decides autonomously with human review only by exception (Human-in-the-Command or fully automated) |
| **Reversibility** | Outcomes easily reversible with minimal impact (e.g., content suggestions) | Outcomes reversible but may cause inconvenience or delay (e.g., service routing) | Outcomes difficult or impossible to reverse; potential for lasting harm (e.g., credit denial, account closure) |
| **Data Sensitivity** | Uses only aggregated or non-personal data | Uses personal data but not special categories | Uses special category data, biometric data, or infers sensitive attributes |
| **Required Governance** | **Standard documentation**<br>Team-level review<br>Annual monitoring | **Independent validation**<br>Committee approval<br>Quarterly monitoring<br>Fairness testing | **Full model risk management (MRM) validation**<br>Board-level reporting<br>Continuous monitoring<br>Incident playbook<br>Regulatory notification |

*Figure 5.2: AI System Classification Template. Risk-Tiering Framework*

The classification isn't a formula; it's a structured judgment. An AI system that scores "high" on even one factor may warrant high-risk governance. The template gives teams and governance committees a common language for discussing risk and a transparent basis for governance decisions.

• • •

## Documentation Is Not Bureaucracy

I want to address an objection I hear frequently, usually from technology teams: "All this governance and documentation will slow us down. We'll never be able to innovate at the pace we need."

I understand the frustration. But I'd push back, gently but firmly, in two ways.

First, the institutions that move fastest with AI are not the ones with the least governance. They're the ones with the clearest governance. When teams know exactly what's required, what classification their system falls into, what validation is needed, what documentation to produce, and who needs to approve it, they can move through the process efficiently. It's ambiguity that slows things down: not knowing who to ask, not knowing what's expected, not knowing whether something will be blocked later because a step was missed. Clear governance eliminates ambiguity.

Second, in financial services, the alternative to governance isn't faster innovation. It's eventual crisis. An AI model that produces biased lending outcomes, a chatbot that gives harmful financial advice, a fraud detection system that fails: these aren't just reputational risks. They're regulatory risks, legal risks, and in some cases existential risks. The governance investment is insurance against outcomes that could set your AI program back years.

The examiner who walked into the first bank in our opening story didn't see governance as a burden. She saw it as evidence of an institution that took its responsibilities seriously. The CRO who could answer her three questions without hesitation wasn't slowed down by governance. She was empowered by it. She knew her AI systems were sound because the governance had verified it. She could innovate with confidence because the controls were there to catch problems before they became crises.

That's what governance foundations provide: not a cage, but a platform. The stronger the platform, the higher you can build, and the more confidently you can keep the human in the lead.

In the next chapter, we move from governance structures to the human element within them. Because the best policies and frameworks in the world are worthless without people who can exercise meaningful oversight, who can challenge AI decisions, recognize when something is wrong, and intervene when it matters. How do we design oversight that is real, not performative? That's the question Chapter 6 addresses.

CHAPTER 6

# Human Oversight in Practice

Priya Singh reviews two hundred AI-generated credit decisions every day.

She arrives at her desk at eight-fifteen, opens the queue, and starts clicking. Each decision comes with a recommendation, approve or decline, along with a confidence score and a brief summary of the applicant's profile. Priya's job title is Senior Credit Review Officer. Her actual job, most days, is clicking "Confirm."

By eleven o'clock, she's reviewed 112 decisions. She's overridden two. One was a clear data error; the system pulled a deceased applicant's file. The other was a borderline case where she recognized the applicant's employer as a company that had recently announced layoffs, information the model hadn't incorporated. The other 110? She reviewed them in the sense that she looked at them. She read the summaries. She checked that the confidence scores were above the threshold. She clicked "Confirm."

Now here's the question that should keep every financial services leader awake at night: Is Priya performing oversight, or is she performing oversight theater?

This isn't a criticism of Priya. She's diligent, experienced, and genuinely cares about getting decisions right. The problem isn't her competence; it's the design. When you ask a human to review two hundred algorithmically generated decisions in a day, you haven't created oversight. You've created a bottleneck that incentivizes rubber-stamping. You've built a system that looks like human control but functions like automation with an extra click.

And regulators are starting to notice.

• • •

## Beyond the Rubber Stamp

The EU AI Act's Article 14 doesn't just require that a human be present in the decision process. It requires that humans be able to "properly understand the relevant capacities and limitations of the high-risk AI system" and be able to "duly monitor its operation, including in view of detecting and addressing anomalies, dysfunctions and unexpected performance." That's a significantly higher bar than clicking Confirm.

The regulation explicitly acknowledges what researchers call automation bias, the well-documented tendency for humans to defer to machine outputs, especially when those outputs come with confidence scores and technical-sounding explanations. A 2023 study published in *Radiology* demonstrated that even highly trained professionals are susceptible: radiologists interpreting mammograms with AI assistance showed measurable deference to incorrect AI recommendations. If specialists with decades of training struggle to maintain independent judgment when an algorithm agrees with them, imagine what happens to a credit officer reviewing her 150th decision of the day.

This is why the Act goes further, requiring deployers to ensure that human overseers maintain awareness of the "possible tendency of automatically relying on or over-relying on the output produced by a high-risk AI system." In other words, regulators aren't just asking whether a human reviewed the decision. They're asking whether the organizational design made meaningful review actually possible.

The implications for financial services are profound. Every bank, wealth management firm, and insurance company deploying high-risk AI (credit scoring, fraud detection, anti-money laundering, insurance underwriting) needs to answer three questions honestly:

- **Can your reviewers actually understand** what the AI system is doing, why it reached a particular conclusion, and what its limitations are?
- **Can your reviewers realistically challenge** the AI's output given their workload, time constraints, and organizational incentives?
- **Can your reviewers effectively intervene** when they identify a problem, and does the organization treat that intervention as valuable rather than disruptive?

If the honest answer to any of these is no, then what you have isn't oversight. It's a compliance checkbox.

• • •

## The Attention Budget: Designing for Human Limits

Here's an uncomfortable truth that most governance frameworks ignore human attention is finite, and it's expensive. You cannot meaningfully review every AI-generated decision. If you try, you'll either create Priya's situation, where the sheer volume makes genuine scrutiny impossible, or you'll slow your operations to the point where the AI system provides no efficiency benefit at all.

The solution isn't to choose between oversight and speed. It's to design systems that spend human attention where it matters most. I call this the attention budget, and it's one of the most important design concepts in responsible AI deployment.

Think of it this way. Your organization has a limited pool of qualified reviewers with limited hours in the day. Every decision you route to them for manual review costs attention, attention that can't be spent on another decision. If you spend that attention on routine, high-confidence decisions where the AI is almost certainly correct, you've wasted your most valuable oversight resource. If you spend it on the genuinely ambiguous, high-stakes, or unusual cases, you've invested it where human judgment adds the most value.

The practical application is a tiered review framework built around three levels of human involvement:

- **Automated with monitoring.** Low-risk, high-confidence decisions proceed without individual human review. Instead, oversight is statistical: a random sample is reviewed daily, and aggregate patterns are monitored for drift, bias, or anomalies. Example: a fraud detection system that clears routine transactions as legitimate with 99.7 percent confidence. No human reviews each individual clearance, but an analyst reviews daily summaries and investigates any shifts in false-positive rates.
- **Human-in-the-Command.** Medium-risk or moderate-confidence decisions proceed automatically but are flagged for review within a defined time window. The human can reverse or modify the decision after the fact. Example: an AI system that adjusts a customer's credit limit based on spending patterns. The adjustment takes effect immediately, but a credit analyst reviews all adjustments above a certain threshold within 48 hours and can reverse any that don't meet policy criteria.
- **Human-in-the-Loop.** High-risk or low-confidence decisions require human approval before they take effect. The AI provides a recommendation and supporting analysis, but no action occurs until a qualified human makes the call. Example: a mortgage underwriting system that recommends denial for an applicant with an unusual employment history. The recommendation goes to an underwriter who reviews the full file, considers context the model may have missed, and makes the final decision.

The key is that these tiers aren't static. They should be calibrated based on the specific risk profile of each decision type, the AI system's demonstrated accuracy and reliability, the downstream impact on customers, and the regulatory requirements for that particular use case. A system that processes insurance claims for minor amounts requires a different oversight model than one that determines whether a customer qualifies for a mortgage.

• • •

## Designing Intervention Points

Getting the tiers right is only half the challenge. The other half is making intervention actually work when it's needed. This is where most organizations fail, not because they don't give humans the authority to override, but because they don't design the experience of overriding to be practical, safe, and valued.

Let me tell you about Daniel, a wealth management advisor at a mid-tier bank. Daniel's firm recently deployed an AI system that generates personalized portfolio rebalancing recommendations for clients. Daniel can accept or modify the recommendations before they go to clients. In theory, he has full override authority.

In practice, here's what happens. Daniel opens his queue and sees thirty-seven recommendations for the day. Each one includes a detailed analysis of the client's risk profile, market conditions, and suggested trades. To meaningfully evaluate a single recommendation, Daniel would need to review the client's full portfolio context, assess whether the AI's risk assumptions are current, consider any personal circumstances the model doesn't know about, and compare the recommendation against his own professional judgment. That takes fifteen to twenty minutes per client. For thirty-seven clients, that's nine to twelve hours of review, before he does anything else in his day.

So Daniel does what any rational person would do. He scans the recommendations, spots the ones that look obviously wrong or unusual, spends real time on those, and approves the rest with a quick glance. He's not lazy. He's trapped in a system that was designed to provide override authority without providing the time, tools, or incentives to actually use it.

Effective intervention design requires addressing three dimensions:

**The challenge function.** It must be organizationally safe to disagree with the AI. If overriding a recommendation triggers extra documentation, manager review, or implicit signals that you're "slowing things down," people will stop overriding. The best organizations I've worked with treat overrides as valuable data: each one is an opportunity to improve the model and understand its limitations. They track override rates not as a problem to minimize but as a signal of healthy human judgment.

**Pre-decision versus post-decision oversight.** Not every decision needs to be caught before it reaches the customer. For some use cases, post-decision review within a defined window is more practical and equally effective. The key is matching the oversight timing to the reversibility and impact of the decision. A product recommendation that a customer hasn't acted on yet can be corrected after the fact. A loan denial that triggers a regulatory reporting obligation cannot.

**Escalation without friction.** When a reviewer identifies a concern, the path to escalation should be immediate, clear, and low-effort. If raising a flag requires filling out a three-page form, scheduling a meeting, or explaining yourself to multiple layers of management, the concern won't get raised. The best escalation systems I've seen use a single-click flag that routes to a dedicated triage function, with the reviewer able to add context in plain language rather than through structured forms.

• • •

## Explainability and Traceability for Reviewers

Here's something that gets lost in most discussions about AI explainability: the question isn't whether the AI can explain itself. The question is whether it can explain itself to the right person, in the right way, at the right moment.

A credit model's internal logic (feature weights, decision boundaries, interaction effects) is useful to the data science team that built it. It's useless to the credit officer reviewing a denial recommendation at nine-thirty on a Tuesday morning. What that officer needs to know is: Why is the system recommending denial? What are the three most influential factors? Has this applicant's profile changed recently? Are there similar cases where the model was wrong?

I think of this as audience-calibrated explainability, and it requires building different explanation layers for different roles:

- **For frontline reviewers** (advisors, underwriters, claims adjusters): plain-language summaries of the key factors driving the recommendation, with clear indicators of confidence level and known limitations. Think of this as the "explain like I'm a relationship manager" layer.
- **For risk and compliance officers**: more detailed views that include the model's performance metrics for this type of decision, historical accuracy on similar profiles, and any flags for potential bias or fairness concerns. This is the "explain like I'm an auditor" layer.
- **For technical teams** (model validators, data scientists): full access to feature importance, model internals, training data characteristics, and drift metrics. This is the "explain like I'm a data scientist" layer.

Each layer serves a different oversight function. Frontline reviewers are catching individual errors and applying contextual judgment the model lacks. Risk officers are monitoring systemic patterns and regulatory compliance. Technical teams are ensuring the model itself remains sound. All three are essential, and none of them can do their job if the explanations they receive aren't tailored to their expertise and decision context.

Traceability is the other half of this equation. For every AI-generated decision, your organization should be able to trace backward from the output to the data inputs, model version, and logic that produced it. When a regulator asks, "Why did your system make this recommendation for this customer on this date?"

the answer cannot be "We're not sure; the model has been updated since then." Financial regulators (the ECB, the Fed, the FCA) are increasingly expecting this level of traceability as a baseline. The US banking regulators' Model Risk Management guidance, originally published in 2011, refreshed in 2026, and now being applied with renewed vigor to AI systems, explicitly requires institutions to maintain robust documentation throughout the model's lifecycle, including clear records of model purpose, limitations, and validation results.

• • •

## Training Staff to Oversee AI Effectively

You can design the perfect oversight system (tiered reviews, intervention points, calibrated explanations), and it will still fail if the people operating it don't have the skills, knowledge, and confidence to use it well.

This is the training gap that almost every financial institution I've worked with underestimates. Organizations invest millions in AI systems and governance frameworks, then give their frontline reviewers a two-hour training session and a user manual. The result is predictable: reviewers who know how to click the buttons but don't know how to evaluate what the system is telling them.

Effective AI oversight training covers four capabilities:

**Recognizing unreliable output.** AI systems fail in patterns. Large language models hallucinate. Credit models miscalibrate confidence on edge cases. Fraud detection systems generate false positives in waves when market conditions shift. Reviewers need to know these patterns, not at the level of a data scientist, but at the level of a skilled practitioner who knows what "normal" looks like and can spot when something is off. This is analogous to how experienced pilots learn to recognize instrument malfunctions: not by understanding the electronics, but by noticing when the readings don't match what they're seeing out the window.

**Recognizing unfair output.** Bias in AI systems often isn't obvious. It shows up as subtle patterns: a credit model that performs well overall but systematically underestimates the creditworthiness of applicants from certain geographic areas. A claims processing system that takes longer to approve claims from older customers. Reviewers need to develop an instinct for these patterns and the knowledge to raise concerns when something feels systematically wrong, even if each individual decision looks reasonable in isolation.

**Confident escalation.** Perhaps the most important capability is the willingness and ability to raise concerns without fear of being seen as disruptive or technophobic. This is as much a cultural challenge as a training challenge. Organizations need to create an environment where questioning the AI is rewarded, not punished, where the reviewer who flags a potential bias pattern is recognized as adding value, not creating problems. Training should include realistic scenarios where escalation is the right call, reinforcing that the human's judgment is valued precisely because it catches things the model cannot.

**Calibration exercises.** Just as financial auditors maintain their skills through regular practice, AI reviewers need ongoing calibration. This means regular exercises where reviewers compare their independent judgment against AI output, review cases where the AI was known to be wrong, and discuss borderline cases as a team. The goal isn't to make reviewers agree with each other; it's to keep their independent judgment sharp so they don't drift into passive confirmation of whatever the system recommends.

One institution I worked with implemented what they called "oversight drills," quarterly exercises where reviewers received a batch of AI decisions that included deliberately planted errors (with management's knowledge, not the reviewers'). The exercise measured how many planted errors each reviewer caught, and the results were used not for performance evaluation but for targeted coaching. Within two quarters, error detection rates improved by over 40 percent. The investment in training paid for itself many times over in avoided regulatory issues and customer complaints.

• • •

## Framework: Oversight Design Canvas

The following canvas provides a template for mapping each AI decision type to an appropriate oversight model. For every AI system in your organization, complete one row. The goal is to ensure that oversight is proportional, practical, and clearly assigned.

| Decision Type | Risk Level | Oversight Tier | Reviewer Role | Escalation Path | Review Frequency |
|---|---|---|---|---|---|
| Routine fraud screen (< $500 transaction) | Low | Automated with monitoring | Fraud analyst (aggregate review) | Anomaly alert → Team lead → Fraud manager | Daily sample + weekly pattern review |
| Credit limit adjustment (existing customer) | Medium | Human-in-the-Command | Credit analyst | Policy exception → Senior analyst → Credit committee | All adjustments > threshold within 48 hrs |
| Mortgage application decision | High | Human-in-the-Loop | Senior underwriter | Decline → Second review → Regional credit officer | Every application before decision issued |
| AML suspicious activity report filing | Critical | Human-in-the-Loop (dual review) | AML analyst + BSA officer | Any SAR → BSA officer → Compliance director | Every filing before submission |
| *[Your decision type]* | *[L / M / H / C]* | *[Tier]* | *[Role]* | *[Path]* | *[Frequency]* |

*Figure 6.1: Oversight Design Canvas*

Instructions: Complete one row for each AI-enabled decision type in your organization. Review the canvas quarterly and update as systems evolve, accuracy data accumulates, and regulatory expectations change. The goal is not to fill this in once; it's to maintain it as a living governance document.

• • •

## Framework: Reviewer Training Curriculum Outline

The following four-module curriculum provides a foundation for building AI oversight capability across your organization. Each module can be delivered as a half-day workshop or adapted into an e-learning format. The curriculum is designed for frontline reviewers and first-line risk staff; separate, deeper programs should be developed for model validators and technical teams.

| Module | Focus Area | Key Learning Outcomes | Suggested Activities |
|---|---|---|---|
| **Module 1** | Understanding AI Fundamentals | Participants can explain how the AI systems they oversee generate recommendations; identify key inputs, outputs, and confidence indicators; and articulate the system's known limitations without requiring technical depth. | Interactive walkthrough of one live system; "Ask the Data Scientist" Q&A session; quiz on system inputs and outputs. |
| **Module 2** | Spotting Errors and Anomalies | Participants can recognize common failure patterns (hallucination, confidence miscalibration, data staleness); identify when output doesn't match contextual knowledge; and apply a structured checklist for evaluating flagged decisions. | Review batch of past decisions including known errors; "Find the Flaw" group exercise; develop personal checklist for daily reviews. |

| Module | Focus Area | Key Learning Outcomes | Suggested Activities |
|---|---|---|---|
| **Module 3** | Detecting Bias and Fairness Issues | Participants can identify patterns suggesting systematic bias (geographic, demographic, temporal); understand the difference between individual fairness and group fairness; and know when and how to escalate bias concerns. | Case studies of real-world AI bias incidents in financial services; pattern recognition exercise with synthetic data; role-play escalation scenarios. |
| **Module 4** | Confident Escalation and Override | Participants can use the escalation pathway effectively; document overrides with sufficient context for downstream review; and contribute to model improvement by providing structured feedback on system performance. | Simulated override scenarios with time pressure; practice documenting overrides; group debrief on escalation culture and psychological safety. |

*Figure 6.2: Reviewer Training Curriculum Outline*

Delivery recommendation: Run Module 1 as a prerequisite before system go-live. Deliver Modules 2 through 4 within the first quarter of deployment. Repeat Module 2 quarterly as a calibration exercise, incorporating recent cases and updated error patterns. Track override rates, error detection rates, and escalation frequency as leading indicators of oversight quality.

• • •

## From Oversight to Understanding

Let me return to Priya, our credit review officer from the beginning of this chapter. Imagine her organization redesigned her role using the principles we've discussed. Instead of reviewing two hundred decisions, she reviews twenty, the ones the system flagged as low-confidence, unusual, or potentially sensitive. Each one comes with a plain-language explanation of why it was flagged and what the key risk factors are. She has fifteen minutes per case instead of ninety seconds. She's been through the four-module training program and participated in quarterly calibration exercises. When she overrides a recommendation, it's recorded as valuable feedback and routed to the model improvement team.

Is Priya still performing oversight? Absolutely. But now it's real oversight, not theater. She's adding genuine human judgment to the decisions where it matters most, and the organization is getting better data about where its AI systems succeed and where they fall short.

That's the transformation this chapter is about. Not whether to have human oversight; that's not negotiable, legally or ethically. But how to design oversight that is meaningful, sustainable, and proportionate. How to spend your attention budget wisely. How to build intervention points that people actually use. How to explain AI output in ways that empower reviewers rather than overwhelm them. And how to train the people who stand between the algorithm and the customer to do their job with skill, confidence, and the full support of their organization.

Effective oversight isn't a constraint on AI adoption. It's what makes AI adoption trustworthy. It's the difference between an organization that deploys AI and hopes for the best, and one that deploys AI and knows (because human beings are genuinely watching, understanding, and intervening when it matters) that it's working the way it should. That is the human in the lead, not a rubber stamp, but a genuine mind engaged with the machine.

> *Oversight depends on people who understand what they're overseeing. Chapter 7 addresses the foundational capability that makes everything else possible: AI fluency across your organization.*

CHAPTER 7

# Building AI Fluency Across the Organization

The question arrives about forty minutes into the quarterly board meeting, right between the risk report and the technology budget review.

A non-executive director (experienced, well-intentioned, widely respected) leans forward and asks: "Can't we just use ChatGPT for everything?"

The room goes quiet for half a second. The Chief Technology Officer glances at the Chief Risk Officer. The CEO looks at her notes. The non-executive director isn't being flippant. He's read the coverage. He's seen the demos. His grandchildren use it for homework. He's asking a reasonable question based on the information available to him, which is the information available to most people outside the technical core of any organization.

But the question reveals something important. Not ignorance, curiosity. And not just curiosity, a gap. A gap between what the board needs to understand about AI to govern it effectively, and what they actually understand today. That gap is what this chapter is about.

Here's why it matters. When board members lack the fluency to ask sharp questions about AI, they default to one of two positions: uncritical enthusiasm ("This sounds transformative, let's invest") or blanket caution ("This sounds risky, let's wait"). Neither position leads to good governance. Good governance requires the ability to interrogate, to ask questions like: What training data was used? How do we know this model isn't biased against certain customer segments? What happens when the model is wrong? Who's accountable? These aren't technical questions. They're governance questions. But asking them requires a baseline level of AI fluency that most boards, most leadership teams, and most organizations simply don't have yet.

Recent workforce surveys find that while the overwhelming majority of leaders agree their company needs AI to stay competitive, most worry that their organization lacks a plan and vision to implement it. Meanwhile, large shares of employees expect their roles to change significantly within the next year, yet only a small minority use AI frequently today. Many feel unprepared for AI-driven changes, and many say their employer expects them to learn AI on their own.

Those numbers tell a clear story: the AI fluency gap isn't a future problem. It's a current one. And in financial services (where AI decisions affect people's mortgages, insurance claims, retirement savings, and fraud alerts), the consequences of that gap are not abstract.

• • •

## What AI Fluency Looks Like (and Doesn't)

Let's start by clearing up the most common misconception. AI fluency is not "everyone learns to code." It's not "everyone becomes a data scientist." It's not even "everyone understands how neural networks work."

AI fluency is contextual understanding. It's the ability, calibrated to your role, to understand what AI systems can and cannot do, how to evaluate their outputs, and how to ask the right questions when something doesn't look right. It's the credit officer who knows enough about how the scoring model works to recognize when a recommendation feels off. It's the branch manager who understands that the customer engagement tool uses historical patterns, not crystal balls, and can explain that to a concerned client. It's the board member who, instead of asking "Can't we just use ChatGPT?" asks "What are the failure modes of the AI systems we've deployed, and how are we monitoring for them?"

The critical insight is that different roles need different depths of fluency. Asking your board to understand gradient descent is as misguided as asking your data scientists to master regulatory capital requirements at CFO depth. Each role has a specific fluency need, and an effective program meets people where they are.

I find it helpful to think about fluency in four tiers:

- **Awareness.** Understanding that AI exists in your organization, what it's being used for, and why it matters. This is the baseline for every employee. If someone in your organization doesn't know which of their daily tools are AI-powered, your awareness tier has failed.
- **Understanding.** Grasping how AI systems work at a conceptual level, not the mathematics, but the logic. What does it mean that a model was "trained on historical data"? Why might that create bias? What's the difference between a recommendation and a prediction? This tier is essential for anyone who uses or reviews AI outputs.
- **Application.** The ability to work effectively with AI tools, evaluate their outputs critically, and integrate AI into workflows while maintaining human judgment. This is the tier for frontline staff, middle managers, and operational leaders who interact with AI systems daily.
- **Leadership.** The ability to make strategic decisions about AI deployment, govern AI systems effectively, and shape organizational culture around responsible AI use. This tier is for executives, board members, and senior risk officers who set the direction.

Notice that each tier builds on the previous one, but the knowledge at each level is qualitatively different, not just more of the same. A board member at the Leadership tier doesn't need to know everything a frontline user at the Application tier knows; they need different knowledge, oriented toward governance, strategy, and accountability rather than daily tool use.

• • •

## Tailored Fluency Programs

Once you accept that fluency is role-dependent, the next question is practical: what does each audience actually need to learn, and how do you deliver it?

**Board and non-executive directors.** This is often the hardest audience because they have the least time and the most influence. A board member who misunderstands AI capabilities can steer an entire organization in the wrong direction, either by greenlighting reckless deployment or by blocking beneficial innovation out of unfounded fear.

What boards need is what I call AI 101 for governance: a concise, jargon-free foundation that enables them to fulfill their oversight role. This means understanding the basics of how AI systems make decisions, knowing the key risk categories (bias, explainability, drift, security), being familiar with the regulatory landscape (especially the EU AI Act's requirements for human oversight and risk management), and, critically, knowing what questions to ask management.

The delivery model matters as much as the content. Annual two-hour board briefings are insufficient for a technology that evolves quarterly. Best practice is a combination of structured onboarding for new directors (a half-day session covering AI fundamentals and the organization's specific AI portfolio), quarterly updates integrated into existing board meetings (fifteen-minute briefings on new deployments, performance metrics, and emerging risks), and access to independent expertise (an external AI advisor who can provide the board with a perspective unconstrained by management's enthusiasm for its own projects).

**Executive team.** Executives need strategic fluency, deeper than the board's governance view, but still oriented toward decision-making rather than technical implementation. The executive who understands that an AI model's accuracy on training data doesn't guarantee accuracy in production, that deployment speed must be balanced against governance readiness, and that the ROI of AI includes risk-adjusted costs for monitoring, oversight, and potential remediation, that executive makes fundamentally better decisions than one operating on demos and vendor promises.

For executives, fluency programs should be scenario-driven. Instead of teaching them what a confusion matrix is, walk them through a scenario: "Your fraud detection model is catching 94 percent of actual fraud. It's also flagging 8 percent of legitimate transactions as suspicious. Customer complaints are rising. What do you do?" That scenario teaches the same concepts (precision, recall, the trade-off between false positives and false negatives) but in a language and context that maps directly to executive decision-making.

**Middle management.** This is the most underserved audience in most AI fluency programs, and it's arguably the most critical. Middle managers are the bridge between strategy and execution. They manage the teams that use AI tools daily. They're the ones who notice when something isn't working, who hear frontline concerns, and who translate executive directives into operational reality.

Middle managers need operational fluency: how to manage AI-augmented teams, how to interpret AI outputs well enough to coach their staff, how to recognize when the AI is creating problems rather than solving them, and how to escalate effectively when they spot issues. They also need the human skills of change management, how to address the anxiety that comes with AI adoption, how to reframe roles so that staff see AI as augmentation rather than replacement, and how to maintain team morale during a transition that many people experience as threatening.

**Frontline staff.** For the people who interact with AI systems every day (advisors, underwriters, claims processors, customer service representatives, compliance analysts), fluency is intensely practical. They need to know how to use the tools effectively, how to recognize when the AI is wrong, and how to maintain their independent professional judgment in a world where algorithms are constantly nudging them toward predetermined conclusions.

The biggest risk at the frontline level isn't that staff won't use AI; it's that they'll use it uncritically. Automation bias, which we explored in Chapter 6, is a frontline phenomenon. The advisor who trusts the portfolio recommendation without checking it against what she knows about the client. The underwriter who accepts the risk score without considering the context the model doesn't see. Frontline fluency training must explicitly address this tendency and build the habit of critical evaluation into daily workflows.

• • •

## Use-Case-Based Learning: Start with Quick Wins

Abstract AI training fails. I've seen it fail repeatedly across multiple institutions. You bring in an expert, they give a two-hour presentation on machine learning concepts, people nod politely, and three weeks later nobody remembers anything because none of it connected to their actual work.

What works is use-case-based learning, teaching through real scenarios that people encounter in their jobs. And the smartest way to start is with quick wins: AI applications that deliver visible, immediate value with low risk and minimal disruption.

Here's why quick wins matter so much for fluency. When you show a relationship manager that an AI tool can draft a first version of client meeting notes in thirty seconds, saving two hours per week of administrative work, you've done three things simultaneously. First, you've created a tangible benefit that makes the person's life better today. Second, you've demonstrated what AI can actually do in a context they understand. Third, and most importantly, you've lowered resistance. The next conversation (about using AI for more complex tasks like portfolio analysis or risk assessment) starts from a position of trust rather than fear.

The sequence matters. Start with AI applications that are:

- **Low risk, high visibility.** Meeting notes, report drafting, data summarization, scheduling optimization. These applications don't make decisions that affect customers; they free up human time for higher-value work.
- **Personally beneficial.** Staff need to feel that AI is helping them, not surveilling them. The first AI experience should save time, reduce tedium, or make a task easier, not add a new monitoring layer to their performance.
- **Easy to evaluate.** People should be able to immediately see whether the AI output is good or bad. Meeting notes are easy to evaluate; you were in the meeting. This builds the habit of critical evaluation before the stakes get higher.

Once quick wins have established a foundation of familiarity and trust, you can progress to more complex use cases: AI-assisted customer interactions, automated compliance checks, predictive analytics for risk management. Each step builds on the fluency developed in the previous one.

Peer champions accelerate this process dramatically. In every organization, there are early adopters, people who are naturally curious about new tools and comfortable experimenting. Identifying these individuals and empowering them as AI ambassadors creates a network of informal coaches who can support their colleagues in a way that formal training programs can't. The advisor who figured out a clever way to use the AI research tool to prepare for client meetings is more credible and more accessible than any training video. Peer learning works because it's contextual, immediate, and grounded in shared experience.

• • •

## Ethics, Bias, and Governance in Every Fluency Effort

Here's where most AI fluency programs make their biggest mistake: they treat ethics and governance as a separate module, something bolted on at the end, after the "real" training about how to use the tools.

This is backward and dangerous. A person who knows how to use AI but doesn't know how to question it is not AI-fluent; they're AI-dependent. And in financial services, AI dependency without ethical awareness creates risks that can cascade through the organization.

Consider this scenario. A team of analysts uses a generative AI tool to draft regulatory reports. They're fluent in the tool; they know the prompts, they know the output format, they're fast and productive. But they haven't been trained to check whether the AI is introducing subtle inaccuracies, paraphrasing regulatory requirements in ways that subtly change their meaning, or using language that could be misinterpreted by regulators. The reports look professional and arrive on time. But they contain errors that a human expert would catch, if anyone were actually reading them critically. This isn't a hypothetical. It's a pattern I've seen emerging across multiple institutions as generative AI tools are adopted for compliance and reporting functions.

Ethics and governance must be woven into fluency from day one, at every tier:

- **At the Awareness tier:** Everyone should understand that AI systems can be biased, can make mistakes, and require human oversight. This isn't about making people afraid of AI; it's about establishing healthy skepticism as a baseline.
- **At the Understanding tier:** People should grasp why bias happens (historical data reflects historical inequities), why explainability matters (you can't govern what you can't understand), and why human oversight isn't optional (both ethically and legally).
- **At the Application tier:** Frontline staff should know how to spot potential bias in the outputs they review, how to escalate concerns, and what their specific governance responsibilities are. They should understand that "the AI

said so" is never a sufficient justification for a decision that affects a customer.

- **At the Leadership tier:** Executives and board members should understand their personal accountability under emerging regulation, the reputational risks of biased AI, and the organizational structures needed to ensure ongoing ethical oversight.

There's one more dimension that most fluency programs miss entirely: the pace of change. AI capabilities are evolving so rapidly that any fluency program built as a one-time event will be obsolete within months. The model your staff learned about in January may have been updated, retrained, or replaced by June. The regulatory landscape that your board was briefed on in Q1 may have shifted substantially by Q3.

This means fluency programs must be living systems, continuously updated, regularly refreshed, and embedded into the rhythm of the organization rather than treated as a project with a completion date. Quarterly refreshers, monthly newsletters, dedicated Slack channels for AI updates, lunch-and-learn sessions featuring recent case studies: these ongoing touchpoints keep fluency current and top of mind.

• • •

## Framework: The Four Tiers of AI Fluency

The following model maps four tiers of AI fluency to specific roles, learning objectives, and delivery methods. Use this framework to assess your organization's current fluency state and design targeted programs for each audience.

| Tier | Target Audience | Learning Objectives | Recommended Delivery |
|---|---|---|---|
| **1. Awareness** | All employees | Identify AI-powered tools used in daily work; explain at a high level what AI does and doesn't do; understand why human oversight matters; know where to report concerns about AI output. | 30-minute e-learning module at onboarding; annual refresher; organizational communications (intranet, newsletters). |
| **2. Understanding** | Staff who use or review AI outputs (advisors, analysts, processors, underwriters) | Explain how AI models learn from data and why that creates bias risk; interpret confidence scores and key output indicators; recognize common failure patterns; understand basic fairness and explainability concepts. | Half-day workshop (instructor-led or virtual); role-specific scenarios and case studies; quarterly calibration exercises. |
| **3. Application** | Middle managers, team leads, operational supervisors | Manage AI-augmented teams effectively; coach staff on critical evaluation of AI output; interpret AI performance metrics; escalate systemic concerns to governance functions; lead change management through AI transitions. | Two-day program with simulations; peer learning cohorts; ongoing community of practice; access to AI governance contacts. |

| Tier | Target Audience | Learning Objectives | Recommended Delivery |
|---|---|---|---|
| **4. Leadership** | Executives, board members, senior risk and compliance officers | Evaluate AI investment proposals critically; govern AI systems with appropriate challenge; understand regulatory requirements (EU AI Act, model risk management); assess organizational AI readiness; set culture and accountability for responsible AI. | Structured onboarding for new leaders; quarterly board briefings; annual strategy session; independent external advisory access. |

*Figure 7.1: AI Fluency Framework. Four-Tier Model*

• • •

## Framework: 90-Day AI Fluency Launch Plan

The following phased rollout plan provides a practical starting point for introducing AI fluency across your organization. Adjust timelines based on organizational size and complexity, but maintain the sequencing: leadership first, then broad awareness, then targeted depth.

| Phase | Timeline | Activities | Success Indicators |
|---|---|---|---|
| **Phase 1 Foundation** | **Days 1–30** | Conduct AI fluency baseline assessment across all tiers. Deliver executive and board AI briefing (Tier 4). Appoint AI Fluency Lead and cross-functional working group. Inventory all AI systems currently deployed or in development. Draft role-specific fluency requirements. | Baseline assessment completed; executive briefing delivered with documented Q&A; AI inventory finalized; fluency lead appointed. |

| Phase | Timeline | Activities | Success Indicators |
|---|---|---|---|
| **Phase 2 Broad Reach** | **Days 31–60** | Launch Tier 1 Awareness module for all employees. Identify and recruit peer champions (target: 1 per 50 employees). Begin Tier 2 Understanding workshops for frontline staff in highest-impact AI areas. Deploy first quick-win AI use case with accompanying training. Establish AI fluency communication channels (newsletter, intranet page, Q&A forum). | 80%+ completion of Awareness module; peer champions identified and briefed; first Understanding workshops delivered; quick-win use case live with positive feedback. |
| **Phase 3 Depth** | **Days 61–90** | Launch Tier 3 Application program for middle managers. Conduct first calibration exercise (AI oversight drill) for Tier 2 staff. Deliver second quick-win use case. Establish quarterly fluency refresh calendar. Measure post-training fluency scores against baseline. Present 90-day progress report to executive sponsor and board. | Tier 3 program launched; measurable improvement in fluency scores; calibration exercise completed with error detection metrics; refresh calendar established; board progress report delivered. |

*Figure 7.2: 90-Day AI Fluency Launch Plan*

Note: The 90-day plan establishes the foundation. True organizational fluency is a multi-year journey. After the initial launch, transition to a continuous fluency program with quarterly refreshers, annual assessments, and ongoing content updates as AI capabilities and regulatory expectations evolve.

• • •

## From Fluency to Infrastructure

Let me return to our board member and his question about ChatGPT. Imagine his organization has implemented the fluency framework we've outlined. He's been through the Tier 4 leadership program. He's received quarterly briefings. He's had access to an independent AI advisor. Now, instead of asking "Can't we just use ChatGPT for everything?" he asks:

> *"What's our model risk management framework for generative AI, and how does it differ from our approach to traditional predictive models? I've been reading about hallucination rates in large language models. How are we monitoring for that in our customer-facing applications?"*

That's not a different person. It's the same person with different knowledge. And that different knowledge changes every conversation that follows, every investment decision, every risk assessment, every governance discussion. It changes the questions management gets asked, which changes the preparation management does, which changes the rigor of the entire AI program.

That's the power of fluency. It doesn't just make individuals smarter about AI. It makes the entire organization more capable, more rigorous, and more trustworthy, and it is the foundation on which keeping the human in the lead ultimately depends.

But fluent people need the right architecture and operating model to work within. Knowledge without infrastructure is insight without impact. Chapter 8 outlines how to build the technical and organizational infrastructure for people-centric AI at scale.

CHAPTER 8

# People-Centric AI Architecture and Operating Models

Fifteen models. That was the number the Chief Technology Officer kept coming back to.

Her data science team had built fifteen AI models over the past eighteen months. Fraud detection. Customer churn prediction. Document classification. Sentiment analysis for customer complaints. Automated KYC screening. Portfolio risk scoring. Fifteen models, each one impressive in the lab, each one validated against test data, each one presented to the executive team with clean slides and confident projections.

Two made it to production.

One of those, the document classifier, was quietly switched off after three months because nobody had built the monitoring infrastructure to detect when it started misclassifying a new document format that a business unit had introduced. The classifier didn't fail dramatically. It failed silently, misclassifying about 12 percent of incoming documents for six weeks before anyone noticed. By then, the downstream impact (delayed processing, incorrect routing, customer complaints) had eroded whatever goodwill the project had earned.

The CTO understood the problem clearly. It wasn't the models. The data science team was talented. The algorithms were sound. The problem was everything around the models: the infrastructure to deploy them, the monitoring to watch them, the governance to control them, the operating model to sustain them, and, most critically, the architecture that would keep humans meaningfully connected to what the AI was doing.

This is the gap this chapter addresses. Not the intelligence inside AI systems, but the architecture, infrastructure, and organizational design around them. Because in financial services, a brilliant model without the right architecture is like a Formula 1 engine in a car with no steering wheel. It's powerful. It's fast. And it's going to crash.

The numbers bear this out. Gartner reported a 1,445 percent surge in multi-agent system inquiries from the first quarter of 2024 to mid-2025, with predictions that 40 percent of enterprise applications will integrate task-specific AI agents by the end of 2026. Yet surveys keep finding that only about one company in ten has successfully deployed agents in production, and an Infosys study found that just 2 percent of companies met its full standard for AI guardrails in 2025. Organizations are building faster than they can govern, and in financial services, that imbalance has consequences.

• • •

## Orchestrating Agents Around Human Checkpoints

The most significant architectural shift in financial services AI isn't the intelligence of individual models; it's the emergence of multi-agent systems where multiple AI components coordinate to accomplish complex tasks. A customer onboarding process might involve an identity verification agent, a document extraction agent, a risk scoring agent, and a regulatory screening agent, all working together to move an application from submission to decision.

In theory, this is powerful. In practice, it creates a new class of governance challenges. When a single model makes a recommendation, oversight is relatively straightforward: a human reviews the recommendation and decides. When five agents are coordinating across a workflow, the question becomes: Where do humans plug in? At which points does the process pause for human judgment, and at which points does it flow autonomously?

The answer lies in what I call trust boundaries, architectural lines that define where AI can act autonomously, where it needs human approval before proceeding, and where it must stop and wait for human direction. Trust boundaries aren't technical abstractions. They're the practical expression of your organization's risk appetite, regulatory obligations, and ethical commitments, encoded directly into your system architecture.

Designing trust boundaries requires answering three questions for every agent-to-agent handoff in your workflow:

- **What is the reversibility of this action?** If the agent makes a mistake at this step, can it be corrected easily, or does it create downstream consequences that are difficult or impossible to undo? Low-reversibility actions (denying a credit application, filing a suspicious activity report, executing a trade) require human checkpoints before execution.
- **What is the impact magnitude?** A document classification error might cause a minor processing delay. A risk scoring error might cause a customer to be denied a mortgage. The higher the potential impact on customers, the organization, or regulatory standing, the stronger the case for a human checkpoint.
- **What is the system's demonstrated reliability?** A model that has operated in production for two years with well-documented accuracy can earn a wider zone of autonomous action than one deployed last month. Trust boundaries should expand or contract based on accumulated evidence, not assumptions.

The practical implementation looks like this. Your multi-agent workflow proceeds through stages. At defined checkpoints, the system presents a human reviewer with a structured summary: what the agents have done so far, what they're recommending next, the confidence levels involved, and any flags or anomalies detected. The reviewer can approve and let the workflow continue, modify the recommendation and let it continue with adjustments, or halt the workflow and route it to manual processing.

Critically, the architecture must support real-time visibility. Human overseers need dashboards that show what agents are doing across the portfolio, not just individual case reviews, but aggregate views that reveal patterns. Is the fraud detection agent flagging more cases than usual this week? Is the document classifier's confidence score trending downward on a particular document type? These signals are invisible at the individual case level but obvious when you have the right monitoring architecture.

• • •

## Designing for Transparency and Auditability from Day One

Here is the architectural principle that separates organizations that govern AI well from those that don't: transparency and auditability cannot be retrofitted. They must be designed in from the beginning.

I've watched organizations make this mistake repeatedly. They build the AI system first, optimize it for speed and accuracy, deploy it to production, and then, often prompted by an audit finding or a regulatory inquiry, try to bolt on logging, lineage tracking, and explainability after the fact. This never works well. Retrofitting auditability is expensive, disruptive, and invariably incomplete because the original architecture wasn't designed to capture the information that auditors and regulators need.

What does "designed for auditability" look like in practice? Three capabilities are essential:

**Decision logging and lineage.** Every AI-generated decision must be traceable from output back through the complete chain: the data inputs that were used, the model version that processed them, any pre-processing or feature engineering that was applied, the reasoning path the model followed, and the confidence scores associated with the output. This isn't just about storing results; it's about storing the complete context needed to reconstruct why a particular decision was made on a particular day for a particular customer. When a regulator asks, "Why did your system decline this mortgage application on March 15?" you need to be able to answer not just with the output, but with the exact model, data, and logic that produced it.

**Observability infrastructure.** Production AI systems change over time, not because anyone modifies the code, but because the world changes. Customer behavior shifts. Market conditions evolve. Data distributions drift. Observability means continuously monitoring your AI systems for these changes and detecting problems before they reach customers. The key metrics to monitor include prediction accuracy over time (is the model getting worse?), input data distribution (are we seeing data patterns the model wasn't trained on?), fairness metrics across protected groups (is bias emerging or growing?), and confidence calibration (when the model says it's 90 percent confident, is it right 90 percent of the time?).

**Glass-box architecture.** This is the principle that AI system internals should be visible to oversight functions, not just to the data science team that built them. In too many organizations, the data scientists understand how the model works, but the risk team, compliance team, and business stakeholders see only inputs and outputs. A glass-box architecture provides appropriate visibility at each level: detailed technical internals for model validators, structured performance dashboards for risk and compliance, and business-language summaries for executive oversight. The goal isn't to make everyone a data scientist; it's to ensure that the people responsible for governance can actually see what they're governing.

One institution I worked with implemented what they called a "model passport," a standardized document that accompanied every AI model from development through production. The passport included the model's purpose, training data description, known limitations, performance benchmarks, bias assessment results, approved use cases, prohibited use cases, monitoring requirements, and assigned human owners. When an auditor or regulator asked about any model, the passport provided a complete, standardized answer. It took effort to implement, but it transformed the organization's relationship with its regulators from defensive to collaborative.

• • •

## Security and Data Protection as Trust Foundations

In financial services, security isn't a technical requirement; it's a trust requirement. Every customer who provides their financial data to an institution is making an implicit trust agreement: I'm giving you access to some of the most sensitive information about my life, and I trust you to protect it. When AI systems access, process, and act on that data, the trust agreement extends to those systems, and the architecture must reflect that.

Four architectural principles are non-negotiable:

**Behind-the-firewall deployment for sensitive data.** When your AI systems process customer financial data (account balances, transaction histories, credit profiles, investment portfolios), that processing should occur within your organization's security perimeter, not in external cloud environments where data governance is delegated to a third party. This doesn't mean you can't use cloud infrastructure; it means you need clear architectural boundaries around where sensitive data flows. The rush to adopt cloud-hosted large language models has created situations where customer data is being sent to external APIs with insufficient controls around what happens to that data once it leaves the institution's environment.

**Data minimization.** Every AI system should access only the data it needs to perform its specific function, no more. A fraud detection model that analyzes transaction patterns doesn't need access to customer investment portfolios. A chatbot that helps customers check their balance doesn't need access to their complete financial history. Data minimization isn't just a privacy principle; it's a security principle. Every additional data field your AI system accesses is an additional exposure surface. The less data the system touches, the smaller the blast radius if something goes wrong.

**Encryption and access controls.** Data should be encrypted at rest and in transit, with role-based access controls that limit who, and what, can access specific data sets. This extends to the AI systems themselves: a model should have the same access restrictions as a human employee in an equivalent role. If a junior analyst wouldn't have access to a particular data set, neither should an AI agent operating at that level of the workflow.

**Third-party and vendor AI risk.** When you deploy a vendor's AI model (whether it's a cloud provider's language model, a fintech's credit scoring engine, or a regtech firm's compliance tool), you are still accountable for the outcomes. The EU AI Act makes this explicit: deployers of high-risk AI systems bear governance and oversight obligations regardless of who built the model. Your architecture must therefore include clear boundaries around vendor AI: what data flows to the vendor, what the vendor does with it, how the vendor's model is validated and monitored, and what happens when the vendor updates their model without notice.

I once identified $3.2 million in security exposure during a pre-production review of an AI deployment at a major financial institution. The exposure wasn't caused by malicious actors; it was caused by architectural decisions that hadn't been reviewed through a security lens. Data was flowing to an external processing layer without adequate encryption. Access controls hadn't been configured for the AI service account, giving it broader data access than intended. The vendor's model was storing intermediate results in a location that wasn't covered by the institution's data retention policies. None of these were intentional; they were the natural result of building fast without building secure. The fix was straightforward once identified, but it would have been far simpler and cheaper to build the security architecture correctly from the start.

• • •

## When You Didn't Build It: Governing Vendor AI

The $3.2 million exposure I just described had one more detail worth dwelling on: the model at the center of it wasn't built by the institution. It was a vendor's. And that makes it typical, because for most financial institutions, the build-versus-buy question has already been answered by reality. Your core banking platform now ships with AI embedded in it. Your fraud tooling comes from a specialist. Your AML screening is a regtech subscription. Your advisors draft with a cloud provider's language model. Walk the AI estate of a typical mid-sized institution and you'll find that most of it, sometimes nearly all of it, is other people's AI.

This is not a problem to be corrected. Very few institutions should be training their own foundation models, and buying well is often the faster, safer route to capability. The problem is that most governance frameworks (including, if we're honest, most of the published guidance) quietly assume you built the thing. The accountability does not make that assumption. You can outsource the build. You cannot outsource the accountability. The EU AI Act makes this explicit, as we've seen: deployers of high-risk AI carry governance and oversight obligations regardless of who built the model. US model risk guidance has always pointed in the same direction; a vendor model is still your model once it's making decisions about your customers. And no customer who is wrongly declined, and no regulator reviewing the decline, will be moved by the news that the algorithm came from a third party. In their eyes, and in the law's, it's yours.

What makes vendor AI genuinely harder to govern than vendor software is a set of properties that traditional third-party risk management wasn't built for. Opacity: the vendor's model is proprietary, its training data is confidential, and the documentation you receive may be closer to marketing than to a model passport. Silent change: vendors update their models continuously; the system that behaves one way in March may behave differently in June without a single line of your code changing, which means yesterday's validation evidence quietly expires. Benchmark inflation: the accuracy figures in the sales deck were measured on the vendor's data, under the vendor's assumptions, not on your customers in your context. And concentration: when much of the industry leans on the same handful of model providers, an error mode or outage at one of them stops being a vendor issue and starts being a sector issue, a dynamic that regulators, through frameworks like DORA, are increasingly focused on.

None of this argues against buying. It argues for buying like an institution that knows it will own the consequences. In practice, that discipline comes down to five questions, asked before the contract is signed and re-asked for as long as the system runs.

**First, what will the vendor actually show you?** Demand documentation that approaches what you'd require of an internal team: intended use and tested limits, a description of training data at least sufficient to assess relevance and bias risk, fairness testing results, and known failure modes. A vendor who won't tell you what the model is for and where it breaks is asking you to carry risk they won't describe.

**Second, can you validate it on your own data?** Not the vendor's benchmark: your customers, your document formats, your fraud patterns. Insist on a pre-production validation window and the contractual right to re-validate on an ongoing basis. The document classifier from this chapter's opening failed on a format the lab never saw; vendor models fail the same way, just with less warning.

**Third, what happens when they change the model?** This is the clause most institutions miss. Require advance notice of material model updates, the ability to pin or stage versions where the architecture allows it, and a defined re-validation window before updated models touch regulated decisions. A vendor's right to improve their product cannot be allowed to mean an unreviewed model making credit decisions on Tuesday that nobody validated on Monday.

**Fourth, where does your data go?** What flows to the vendor, what they retain, whether your customers' data trains their next model, and what gets deleted, provably so, when the relationship ends. Data minimization applies with extra force here: every field that crosses the boundary is a field you no longer fully control.

**Fifth, can you see, stop, and leave?** Audit rights and regulator access. Decision logs and lineage delivered to you, in your format, not trapped in the vendor's console. A kill switch on your side of the boundary, so that disabling the vendor's AI never depends on the vendor's cooperation or business hours. And a tested exit plan, because a model you cannot leave is a model you cannot govern.

Notice what these five questions really are: the contract working as a governance instrument. Service-level agreements for AI should cover accuracy, drift, and fairness, not just uptime. Incident notification timelines should be measured in hours, not billing cycles. The institutions that get this right don't build a parallel bureaucracy for vendor AI; they extend the machinery they already have (model risk management and third-party risk management, meeting in the middle) and they treat their strategic AI vendors the way they treat any outsourced critical function: with warmth, partnership, and a signed right to verify everything.

As for what's left of build versus buy: build where the capability differentiates you and your data is the advantage; buy where the capability is a commodity and the vendor's scale is the advantage; and in both cases, govern identically. Figure 8.1's principles don't have an asterisk for purchased systems. A vendor model inside your workflow is your model, in the eyes of your customer, your regulator, and this book. The human in the lead doesn't move out when the technology is rented.

• • •

## Operating Model Changes

The best architecture in the world will fail without the right organizational structure to build, operate, and govern it. This is the lesson that our CTO from the chapter's opening learned the hard way: thirteen of her fifteen models failed not because of technical limitations but because the organization's operating model couldn't support them.

Deploying AI at scale in financial services requires fundamental changes to how teams are structured, how decisions are made, and how accountability flows. Four structural elements are essential:

**Cross-functional AI councils.** AI governance cannot live within a single function. A model that affects customers involves technology, business, risk, compliance, legal, and customer experience. An AI council brings these perspectives together, not as a one-time review gate, but as an ongoing governance body that oversees the organization's AI portfolio. The council sets standards, reviews high-risk deployments, resolves cross-functional disputes, and provides a single point of accountability for the board and regulators. The most effective councils I've seen meet monthly, maintain a portfolio dashboard of all active AI systems, and operate with clear delegated authority from the board.

**Product squads for AI.** The traditional model (where data scientists build models and "throw them over the wall" to engineering for deployment) doesn't work for AI that needs to operate reliably in production. The alternative is multidisciplinary product squads: small, dedicated teams that include data scientists, software engineers, business analysts, and embedded risk partners. These squads own the AI product end-to-end, from development through deployment through ongoing monitoring and improvement. They're responsible not just for building the model but for keeping it running, keeping it accurate, and keeping it governed.

**Risk partners embedded in delivery.** In traditional technology delivery, risk and compliance review happens at the end, a gate that the project must pass through before going live. For AI, this model is dangerously inadequate. By the time a model reaches the review gate, architectural decisions have already been made that are expensive or impossible to reverse. Embedding risk partners directly into AI delivery teams transforms the relationship from adversarial ("risk says no") to collaborative ("risk helps us build it right"). The risk partner participates in design decisions, reviews training data choices, validates fairness assessments, and ensures governance requirements are met incrementally rather than in a last-minute scramble.

**The AI product manager role.** This is an emerging function that bridges the gap between what the business wants, what the technology can do, and what governance requires. An AI product manager doesn't write code or build models. They translate business intent into AI requirements, ensure that governance considerations are baked into the product roadmap, coordinate between the technical team and business stakeholders, and own the ongoing performance and accountability of the AI product in production. In my experience, the absence of this role is one of the most common reasons AI projects fail to move from lab to production. Without someone who owns the entire lifecycle (not just the build, but the operation, governance, and continuous improvement) AI products drift into neglect.

Among organizations already using agentic AI extensively, most expect to change their operating model and redefine roles, including flattening hierarchies and reducing middle management layers. This isn't just about efficiency. It's about recognizing that AI changes how work flows through an organization, and the organizational structure needs to evolve accordingly.

• • •

## Framework: People-Centric Architecture Principles

The following ten principles provide a foundation for designing AI architectures that keep humans meaningfully in control. Use them as a checklist when evaluating new AI system designs or auditing existing deployments.

| # | Principle | Description |
|---|---|---|
| 1 | **Human Checkpoints by Design** | Every AI workflow must include defined points where qualified humans can review, override, or halt the process. These checkpoints must be architectural requirements, not afterthoughts. |
| 2 | **Proportional Autonomy** | The degree of AI autonomy must be proportional to the system's demonstrated reliability, the reversibility of its actions, and the magnitude of potential impact on customers and the organization. |
| 3 | **Complete Decision Lineage** | Every AI-generated output must be traceable from the final decision back through the full chain: data inputs, model version, feature engineering, reasoning path, and confidence scores. |
| 4 | **Audience-Calibrated Transparency** | System explanations must be tailored to the expertise and needs of each stakeholder: business-language summaries for executives, structured dashboards for risk officers, technical detail for model validators. |
| 5 | **Continuous Observability** | Production AI systems must be continuously monitored for accuracy drift, data distribution shifts, fairness degradation, and confidence miscalibration, with automated alerting when metrics breach defined thresholds. |
| 6 | **Data Minimization** | AI systems must access only the data required for their specific function. Broader data access must be explicitly justified, documented, and approved through governance processes. |
| 7 | **Security as Trust Architecture** | Data protection, encryption, and access controls are not technical add-ons; they are trust requirements. AI systems must meet the same security standards as human employees in equivalent roles. |
| 8 | **Vendor Accountability** | Third-party AI models are subject to the same governance, monitoring, and oversight requirements as internally developed systems. Vendor architecture must enable auditability and human intervention. |

| # | Principle | Description |
|---|---|---|
| 9 | **Graceful Degradation** | When AI systems fail, malfunction, or encounter scenarios outside their training data, they must fail safely, routing to human decision-makers rather than producing unreliable outputs or halting operations. |
| 10 | **Living Governance** | Architecture must support ongoing governance, not just launch-day compliance. Trust boundaries, monitoring thresholds, and oversight models must be updatable as the system evolves and as evidence accumulates. |

*Figure 8.1: Ten People-Centric Architecture Principles*

• • •

## Framework: AI Operating Model Canvas

The following canvas provides a template for designing the organizational structure around AI delivery and governance. Complete each cell to map your current state and target state.

| Dimension | Key Questions | Current State | Target State |
|---|---|---|---|
| **Governance Structure** | Who owns the AI portfolio? How are deployment decisions made? Who is accountable to the board and regulators? | *[Assess]* | *[Define]* |
| **Delivery Model** | How are AI products built? Are teams cross-functional? Is risk embedded in delivery or reviewing at the gate? | *[Assess]* | *[Define]* |
| **Monitoring & Operations** | How are production AI systems monitored? Who responds to alerts? How is model drift detected and addressed? | *[Assess]* | *[Define]* |

| Dimension | Key Questions | Current State | Target State |
|---|---|---|---|
| **Talent & Roles** | Do you have AI product managers? Are reviewers trained? Is there an AI fluency program? | *[Assess]* | *[Define]* |
| **Vendor Management** | How are third-party AI solutions evaluated, onboarded, and monitored? Who owns vendor AI risk? | *[Assess]* | *[Define]* |
| **Data Architecture** | Is data minimization enforced? Are access controls applied to AI systems? Is decision lineage captured? | *[Assess]* | *[Define]* |
| **Change Management** | How are affected employees prepared for AI-augmented roles? Who leads cultural change? | *[Assess]* | *[Define]* |

*Figure 8.2: AI Operating Model Canvas*

Instructions: Complete the canvas as a cross-functional exercise involving technology, business, risk, compliance, and HR leadership. The "Current State" column should reflect honest assessment, not aspiration. The "Target State" column should define specific, measurable objectives. Review and update the canvas quarterly as your AI program matures.

• • •

## From Architecture to Action

Let me return to our CTO and her fifteen models. Imagine her organization had built the architecture and operating model we've described before those models were developed. The trust boundaries would have been defined upfront, so each model's zone of autonomy and human checkpoint requirements would have been clear from the design stage. The observability infrastructure would have caught the document classifier's drift within days, not weeks. The embedded risk partners would have flagged the monitoring gap before deployment. The AI product managers would have ensured that each model had a clear owner, a monitoring plan, and a governance framework.

Instead of two out of fifteen making it to production, most of them would have, because the organization would have been ready for them. Not just technically ready, but organizationally ready and governance-ready.

That's the real message of this chapter. The bottleneck in AI adoption isn't intelligence; it's infrastructure. Not just technical infrastructure, but the human and organizational infrastructure that makes people-centric AI safe, governable, and sustainable at scale. Architecture is strategy. Operating models are governance. And both must be designed around the principle that has guided this entire book: humans, meaningfully in the lead.

> *Theory and frameworks only go so far. Chapter 9 brings these ideas to life through stories from the field, what worked, what didn't, and what leaders learned.*

CHAPTER 9

# Case Vignettes from the Field

Every AI project has two stories.

The first is the one that appears in the investor presentation, the internal newsletter, and the technology conference keynote. It's a story of innovation, efficiency gains, and competitive advantage. It follows a clean arc: the organization identified an opportunity, deployed an AI solution, and achieved measurable results.

The second story is messier. It's the one the project team members tell each other over drinks after the launch. It includes the executive sponsor who nearly pulled funding in month four, the compliance team that froze the deployment for six weeks, the operations staff who quietly reverted to manual processes because they didn't trust the model, and the customer who called their branch manager in tears because an algorithm had declined their mortgage application with no explanation.

The second story is where the learning lives.

This chapter tells five of those second stories. Each is a composite, drawn from real experiences across multiple financial institutions, anonymized and combined to protect confidentiality but faithful to the patterns, tensions, and lessons that emerged. Together, they illustrate a recurring theme from every chapter of this book: technology-first AI deployments in financial services almost always require a people-centric course correction before they succeed.

The names are fictional. The problems are not.

• • •

## The Operations Transformation That Almost Failed

*Meridian National Bank • Retail Banking Operations • 18-month program*

The business case was compelling. Meridian National's back-office operations team processed over two million transactions daily: payment exceptions, account reconciliations, trade settlements, and customer document verifications. The work was labor-intensive, repetitive, and expensive. A consulting engagement estimated that AI-powered automation could reduce processing time by 40 percent and deliver annual savings of thirty-two million dollars.

The Chief Operating Officer, David Chen, secured board approval for a phased automation program. The technology team selected a combination of machine learning models for document classification and intelligent process automation for routine transaction handling. The initial pilot targeted payment exception processing, a high-volume, rules-heavy workflow that seemed ideal for automation.

The pilot worked beautifully. In a controlled environment, the AI processed payment exceptions with 94 percent accuracy, better than the team's historical average of 88 percent. Processing time dropped by 60 percent. The executive team was thrilled. David greenlit a rapid rollout across all operations workflows.

That's when the problems started.

Within three months of full deployment, error rates had climbed back to pre-automation levels. Staff satisfaction scores plummeted. Two senior operations managers, people with decades of institutional knowledge, resigned. And a pattern emerged that nobody had anticipated: the AI handled "easy" cases efficiently, but those easy cases had served as the training ground for junior analysts. Without them, new hires had no way to build competence, and the team's ability to handle complex exceptions, the cases the AI couldn't process, deteriorated.

David described the situation bluntly in an internal review: "We automated the floor out from under our own people. The AI was doing the work, but the organization was getting dumber."

The turnaround began when the operations leadership team reframed the initiative. Instead of asking, "What can we automate?" they asked, "How should humans and AI work together in this workflow?" The distinction sounds semantic, but it changed everything.

The redesigned system operated on three principles. First, AI handled initial processing and flagged cases by complexity tier: routine, moderate, and complex. Second, junior analysts worked alongside the AI on moderate cases, reviewing its outputs and learning exception-handling patterns in the process. Third, senior analysts focused exclusively on complex cases and spent a defined portion of their time mentoring juniors and improving the AI's training data.

Career paths were redesigned. Progression was no longer measured by volume processed but by the complexity of cases handled and contributions to AI improvement. A new role emerged, AI operations specialist, for people who bridged domain expertise and model performance.

Twelve months after the redesign, Meridian achieved the efficiency targets from the original business case, but with higher staff satisfaction, lower attrition, and a measurably stronger bench of operations talent. The throughput gains came not from replacing people but from redirecting their expertise to where it mattered most.

***Lesson: The easy cases aren't waste; they're the training ground. Automate thoughtlessly and you hollow out the capability you need for the cases AI can't handle.***

• • •

## The Virtual Assistant That Rebuilt Trust

*Harborview Wealth Management • Digital Client Experience • 24-month initiative*

Harborview Wealth Management launched its virtual assistant, internally called "Aria," with considerable fanfare. The firm managed over forty billion dollars in client assets, and its leadership believed a conversational AI tool could extend the reach of its advisory team, providing instant responses to routine client inquiries and freeing advisors for deeper, relationship-driven conversations.

The initial deployment optimized for a single metric: *containment rate*, the percentage of client interactions resolved entirely within the digital channel without human intervention. The technology vendor had recommended this metric, and it aligned with the CFO's efficiency mandate. Within six months, Aria's containment rate reached 72 percent.

The number looked good on paper. In practice, the firm was losing clients.

The problem surfaced in quarterly client surveys. Net Promoter Scores among digitally active clients dropped by eleven points. Qualitative feedback was scathing. Clients reported that Aria gave "generic, unhelpful responses" to nuanced financial questions. When clients wanted to speak with their advisor, Aria attempted multiple redirections before offering a transfer, because every transfer reduced the containment rate. One client, whose portfolio was worth over eight million dollars, described the experience as "being trapped in a phone tree that didn't want to let me talk to a real person."

The Chief Client Officer, Maria Santos, convened a cross-functional review. What she found was a classic misalignment between the metric being optimized and the outcome that mattered. Containment rate measured whether the AI *kept* the client. It didn't measure whether the AI *helped* the client.

The redesign started with the metrics. Containment rate was replaced by three new measures: *resolution quality* (did the client's issue actually get resolved?), *client effort score* (how easy was the interaction?), and *advisor satisfaction* (did the AI make advisors' jobs easier or harder?). These metrics aligned the AI's behavior with the firm's actual value proposition: trusted, personal wealth management.

Aria's behavior was rebuilt around these new objectives. The AI was trained on transcripts from the firm's highest-rated advisors, not generic financial content, giving it the tone and substance of a Harborview conversation. Seamless human handoff became a feature, not a failure. When Aria detected a complex question, an emotional cue, or a high-value client, it proactively offered to connect the client with their advisor, presenting a summary of the conversation so far so the advisor could pick up without the client repeating themselves.

A new "advisor co-pilot" mode emerged from the redesign. Rather than replacing advisor interactions, Aria prepared advisors for them, surfacing relevant portfolio data, recent client activity, and market context before scheduled calls. Advisors reported that this preparation saved them an average of twenty minutes per client meeting and made conversations noticeably more productive.

Eighteen months after the redesign, Harborview's digital adoption had increased by 35 percent. But the result that mattered most to Maria was this: Net Promoter Scores among digitally active clients were now *higher* than among clients who had never used Aria. The AI wasn't just avoiding harm; it was actively strengthening the client relationship.

***Lesson: Optimize for containment and you contain the client. Optimize for resolution and you build trust. The metric you choose is the behavior you get.***

• • •

## The Fraud System That Balanced Detection with Fairness

*Commonwealth Retail Bank • Fraud Operations • Ongoing program*

Commonwealth Retail Bank's AI-powered fraud detection system was, by most measures, a success. Within its first year, the system had identified 14 percent more fraudulent transactions than the previous rules-based approach, while reducing false positives by 22 percent overall. The fraud operations team was processing fewer alerts, catching more actual fraud, and doing both faster.

Then a data analyst in the consumer advocacy team ran a report that nobody had asked for.

She disaggregated the false positive rates by customer demographics, geography, income level, and account age. The results were troubling. Customers in lower-income zip codes were flagged as potentially fraudulent at nearly three times the rate of customers in affluent areas, after controlling for actual fraud incidence. Customers with accounts less than two years old, disproportionately younger and more diverse, experienced false positive rates 40 percent higher than established customers. The AI's overall accuracy had improved, but its errors were concentrated among specific populations.

The Head of Fraud Operations, James Okafor, initially pushed back. "The model is more accurate overall. That's what matters." But the Chief Risk Officer, who had been reading the emerging regulatory guidance on algorithmic fairness, recognized the risk. A system that was accurate in aggregate but discriminatory in distribution wasn't just an ethical problem; it was a regulatory and reputational liability.

The root cause was familiar to anyone who has studied algorithmic bias. The model had been trained on historical transaction data that reflected existing patterns, patterns shaped by decades of socioeconomic inequality. Transactions in lower-income areas looked different from those in affluent areas: smaller amounts, more cash transactions, more payday-related patterns. The model had learned to treat these legitimate behavioral differences as fraud risk signals.

The remediation involved four interconnected changes.

First, the data science team implemented fairness constraints during model training: equalized odds testing that measured false positive rates across demographic groups and penalized models that showed significant disparities. This didn't mean ignoring fraud patterns in specific populations. It meant ensuring that *innocent* customers in every population had roughly equal chances of being incorrectly flagged.

Second, a human review layer was added for all cases in the top decile of the model's risk scores before any customer-facing action was taken. Analysts in this review role received specific training on confirmation bias, the tendency to look for evidence that confirms the model's judgment rather than evidence that challenges it.

Third, quarterly bias audits became a standing governance requirement. An independent team reviewed disaggregated performance data and reported directly to the risk committee, not to the fraud operations team. This separation of audit from operations was critical; it prevented the team optimizing for detection accuracy from also being the team that judged whether detection was fair.

Fourth, and this was the change that most impressed regulators, Commonwealth created a customer appeal mechanism. Any customer whose transaction was declined or frozen due to a fraud flag could request a human review, conducted within twenty-four hours. The review was conducted by a senior analyst who had access to both the model's reasoning and the customer's full account history. Appeal outcomes were tracked and fed back into model retraining.

James Okafor, the initial skeptic, became the program's strongest advocate after reviewing six months of data. The fraud detection rate had held steady; the fairness constraints hadn't meaningfully reduced the system's ability to catch actual fraud. But customer complaints related to false fraud flags had dropped by 41 percent, and the demographic disparity in false positive rates had narrowed from three-to-one to within 15 percent. "I was wrong to see fairness and accuracy as trade-offs," he told his team. "They're complementary. A system that's unfair is inaccurate; it's just inaccurate in a way that hides behind averages."

***Lesson: Aggregate accuracy can mask concentrated harm. Fairness isn't a constraint on good AI; it's a dimension of it.***

• • •

## The Underwriting Engine That Said No to the Right Customers

*Lakeshore Mutual Life • Accelerated Underwriting • 14-month program*

Lakeshore Mutual's problem was the one every mid-size life insurer knows by heart: the people who most needed its products were the least likely to finish buying them. Traditional underwriting meant medical exams, fluid draws, attending physician statements, and a six-week wait, and middle-market families, the carrier's stated growth priority, abandoned applications at twice the rate of affluent ones. The promise of accelerated underwriting was irresistible. An AI triage model would assess applications using prescription histories, public records, and third-party data, issuing clean cases in days without an exam. The target metric was the industry's favorite: the touchless rate, the share of policies issued straight through, with no human underwriter involved.

The Chief Underwriting Officer, Diane Keller, set the goal at 60 percent touchless within a year. The model cleared it in eight months. The celebration lasted one quarter.

Then the field reports started arriving. Placement rates, applications that actually became policies, hadn't improved. In the middle market, they had fallen. The model was fast at saying yes to the customers who would have sailed through traditional underwriting anyway: established professionals with deep digital footprints, long prescription histories, and stable address records. Confronted with the customers Lakeshore actually wanted to reach (younger families, recent immigrants, gig workers, anyone with a thin file), the model did what models do with uncertainty: it routed them away. Some were referred into the old six-week pathway. Many were quietly rated up or declined. The agents saw it before the dashboards did. "The system gives me an instant yes for the clients who don't need help," one top producer told Keller, "and an instant maybe for everyone I built my book serving." Several began steering business to competitors, not because the prices were better, but because they could promise their clients an answer.

There was a quieter problem underneath. Lakeshore's actuaries, uneasy about mortality risk in a process with no medical evidence, had layered conservative assumptions into the accelerated pathway's pricing. The thin-file customers who survived triage were being charged for the model's uncertainty about them. The carrier was, in effect, penalizing exactly the lack of data history that defined its growth market, and in life and health insurance, where European regulators now classify AI risk assessment as high-risk and US state regulators are sharpening their scrutiny of proxy discrimination, "the model was uncertain" was not going to survive its first market-conduct exam as an explanation.

The redesign reframed the question. Touchless rate measured how often the machine avoided humans. Nobody could explain how that served the customer. The new north-star metric was speed to decision for every applicant, with three pathways instead of a binary. Clean cases still went straight through. Clearly complex cases went to senior underwriters from the start. And the thin-file middle, the cases the model had been quietly turning away, got a new Human-in-the-Command lane: the model assembled what evidence existed, stated explicitly what it didn't know, and routed the file to an underwriter empowered to fill the gap with a phone interview or alternative evidence rather than a decline. Reason codes went to agents in plain language, so a referral stopped feeling like a rejection. Underwriter overrides were logged and fed back into the model's training, and the actuaries, watching early mortality experience track the underwriters' judgments, began unwinding the conservative loads.

Fourteen months in, Lakeshore's touchless rate had settled at 48 percent, twelve points below the original target, a number the old dashboard would have scored as failure. Every other number moved the right way. Median time to decision fell from six weeks to four days for the middle market. Placement rates in the target segment rose by a third. Agent satisfaction with underwriting hit a ten-year high. And when the market-conduct examiners eventually came, Keller could show them something rare: a file-by-file record of why every thin-file applicant got the decision they got, and a human signature on the ones that mattered. "We were optimizing for speed to no-touch," she told her board. "The business needed speed to yes, for the customers we exist to serve."

***Lesson: In insurance, the algorithm doesn't just price risk; it decides who gets protection. A model that handles uncertainty by turning customers away will quietly unwrite your growth strategy. Design for the customer you're trying to reach, not the file you wish they had.***

• • •

## The Board That Got Governance Right (Eventually)

*Pacific Coast Financial Group • Enterprise AI Governance • 12-month transformation*

When regulators sent Pacific Coast Financial Group a formal information request about its AI systems in the spring of 2025, the board discovered something alarming: nobody could answer the questions.

The request was routine, part of a thematic review of AI governance across mid-sized financial institutions. Regulators wanted an inventory of AI systems in production, documentation of oversight mechanisms, evidence of bias testing, and records of model performance monitoring. Standard governance artifacts that any well-run AI program should be able to produce.

Pacific Coast couldn't produce any of them. Not because the firm wasn't using AI; it had over twenty machine learning models deployed across credit scoring, fraud detection, customer segmentation, and marketing optimization. But AI had grown organically. Individual business units had procured or built AI tools to solve specific problems, each with its own vendor, its own data pipeline, and its own informal oversight process. There was no central inventory, no consistent governance framework, and no board-level visibility into what AI systems existed, what they did, or how they were controlled.

The Board Chair, Patricia Reeves, described the realization in stark terms: "We had been governing by assumption. We assumed that because these were technology projects, they were covered by our existing technology risk framework. They weren't. AI systems make decisions that affect customers, and we had no idea what decisions were being made or how."

The remediation began with an inventory. A cross-functional team spent eight weeks cataloging every AI and machine learning system in the organization, not just models built internally, but third-party tools and embedded AI features in vendor products. The inventory revealed thirty-one distinct AI systems, eleven more than anyone had estimated. Seven of them directly influenced customer-facing decisions.

From the inventory, the governance team built a risk-tiered classification system. Systems were classified as high, medium, or low risk based on three criteria from the frameworks described in earlier chapters: customer impact, decision reversibility, and regulatory sensitivity. High-risk systems (credit scoring, fraud detection, and AML screening) received the most intensive governance: quarterly performance reviews, annual bias audits, documented human oversight protocols, and direct reporting to a newly formed AI governance committee.

The AI governance committee was the centerpiece of the new structure. It met monthly, included the Chief Risk Officer, the Chief Technology Officer, the Chief Compliance Officer, and two independent board members. Its charter was straightforward: ensure that every AI system in production had a named business owner, documented performance metrics, tested oversight mechanisms, and a clear escalation path for incidents.

The board itself received quarterly AI risk reports, not technical deep-dives, but structured summaries covering four questions: What AI systems are in production? What decisions are they influencing? What risks have been identified? What actions have been taken? Patricia Reeves insisted on this format because, as she put it, “Boards don’t need to understand gradient descent. They need to understand who is affected and who is accountable.”

Perhaps the most transformative element was the board fluency program. Over three sessions, board members engaged in structured workshops that covered not the mechanics of AI, but the governance implications: how to ask the right questions, what red flags to look for, and how to challenge AI optimism without stifling innovation. The workshops used scenarios drawn from actual regulatory actions and industry incidents, making the learning concrete and relevant.

When regulators returned for a follow-up review nine months later, Pacific Coast not only passed; the examiners cited the firm’s governance framework as “emerging best practice.” More importantly, the internal culture had shifted. Business units that had previously deployed AI informally now proactively engaged the governance process, because they recognized it as enabling rather than blocking. Deployment timelines actually shortened, because governance was now built into the design phase rather than bolted on as a last-minute gate.

"The irony," Patricia noted at a board meeting, "is that governance made us faster. When everyone knows the rules, nobody wastes time negotiating them."

***Lesson: Governance isn't overhead; it's infrastructure. Build it early and it accelerates everything that follows.***

• • •

## The Patterns Beneath the Stories

Five different institutions. Five different use cases: operations, client experience, fraud detection, insurance underwriting, and governance. Yet the same patterns recur across every story.

**The technology worked. The organization wasn't ready.** In each case, the AI itself performed as designed. The failures were organizational: hollowed-out capability at Meridian, misaligned incentives at Harborview, hidden bias at Commonwealth, uncertainty-driven exclusion at Lakeshore, and governance gaps at Pacific Coast. The consistent lesson is that AI readiness is primarily an organizational capability, not a technical one.

**The first deployment was technology-first. The successful redesign was people-first.** Every vignette follows the same arc: deploy with a technology lens, encounter people problems, redesign around human needs, and achieve sustainable results. This isn't a failure pattern; it's a learning pattern. But it's an expensive one. Organizations that start with a people-centric lens can skip the costly middle step.

**Measurement shaped behavior.** Meridian measured volume processed and got hollowed expertise. Harborview measured containment and got trapped clients. Commonwealth measured aggregate accuracy and got concentrated unfairness. Lakeshore measured touchless rate and got turned-away customers. In every case, the redesign began with redefining what success looked like, and the new metrics drove the new behavior. The framework from Chapter 11 provides a structured approach to defining metrics that align AI behavior with organizational values.

**Governance enabled speed, not bureaucracy.** This is the counterintuitive finding that surprised leaders in every story. At Meridian, structured human-AI workflows processed cases faster than either alone. At Harborview, advisor preparation reduced meeting time. At Commonwealth, fairness constraints reduced complaints and rework. At Lakeshore, the three-pathway redesign cut median decision time from six weeks to four days. At Pacific Coast, governance shortened deployment timelines. In each case, the "overhead" of people-centric design delivered returns that exceeded the efficiency gains from technology alone.

**Iteration was essential.** No organization got it right the first time. The difference between success and failure was not the quality of the initial design; it was the willingness to listen to feedback, acknowledge problems honestly, and redesign. Building AI systems that improve over time requires building organizations that learn over time.

• • •

## From Stories to Strategy

These vignettes demonstrate what is possible when financial institutions commit to people-centric AI, not as a philosophy, but as a practice. The operations team that redesigned career paths. The wealth management firm that redefined success metrics. The fraud team that embedded fairness into its pipeline. The insurer that redesigned underwriting around the customers it wanted to reach. The board that built governance infrastructure from scratch. Each story shows that the frameworks and principles described in earlier chapters are not abstractions; they work, in real institutions, with real constraints.

But stories inspire; roadmaps execute. Chapter 10 gives you a concrete plan to make this happen in your organization, starting in the next ninety days.

CHAPTER 10

# The 90-Day People-Centric AI Roadmap

You don't need a three-year strategy to start.

This is the sentence I find myself repeating most often to executive teams who have spent months, sometimes years, debating AI strategy without deploying anything meaningful. They've commissioned consulting studies. They've attended conferences. They've formed committees. And they're no closer to a production AI system that creates value for their customers, their employees, or their shareholders than they were when they started.

The problem isn't a lack of ambition. It's a lack of a concrete, bounded plan that forces decisions, creates evidence, and builds organizational confidence. A three-year roadmap feels safe because it defers hard choices. A ninety-day roadmap is uncomfortable because it demands them.

This chapter provides that ninety-day plan. It's designed for two audiences: leaders launching their first serious AI initiative and leaders resetting an initiative that has stalled. Both situations require the same discipline (alignment, design, execution, and learning) compressed into a timeline that is long enough to produce real evidence and short enough to maintain executive attention and organizational momentum.

The roadmap is organized into three phases. Days one through thirty focus on assessment and alignment, getting the right people in the room, agreeing on principles, and selecting the right pilot. Days thirty-one through sixty focus on design, building the pilot with governance, oversight, and change management baked in from the start. Days sixty-one through ninety focus on execution, measurement, and decision, running the pilot, collecting people-centric evidence, and making an informed choice about what comes next.

Each phase has a specific deliverable. If you complete all three, you'll have something most financial institutions lack: evidence-based confidence about what AI can do in your organization, how your people respond to it, and what it takes to govern it responsibly.

One more thing. The AI systems entering financial services today aren't the rule-based automation tools of a decade ago. They're increasingly capable of reasoning, making decisions, and taking actions with varying degrees of autonomy. Your ninety-day plan must account for this reality. The roadmap that follows includes guidance on how to structure human oversight as these systems grow more capable, a theme we'll explore fully in Chapter 12. For now, the principle is simple: start every AI capability in observe-only mode, and expand its autonomy only when the evidence justifies it.

• • •

## Why Ninety Days?

The ninety-day window is strategically significant for reasons that matter deeply in financial services.

First, **budget cycle alignment**. Financial institutions operate on quarterly reporting cycles. A ninety-day plan maps to a single quarter, giving business and operations heads, the budget holders who must justify the spend, a natural checkpoint to demonstrate return on investment. When you present a plan that fits within a single budget period, you reduce the approvals required and accelerate the path from concept to commitment.

Second, **decision fatigue reduction**. Enterprise decision-makers evaluate dozens of initiatives competing for attention and resources. A concrete ninety-day roadmap with named milestones stands out against vague multi-year strategies that defer every hard choice to a future quarter. Specificity signals competence. Bounded timelines signal confidence.

Third, **risk reduction**. By breaking implementation into three digestible phases, you transform a large, risky technology bet into three smaller, verifiable commitments. Each phase concludes with a gate review, a structured checkpoint where leadership can assess progress, validate assumptions, and decide whether to proceed. No organization is asked to commit fully on day one. Trust is earned in increments.

Fourth, momentum. Ninety days is long enough to produce real evidence and short enough to maintain organizational momentum. The industry's long planning horizons and consensus habits serve risk management well but fit poorly with a technology that evolves faster than any planning cycle. Ninety days forces action.

• • •

## Days 1–30: Assess and Align

The first thirty days are about people, not technology. The most common mistake in AI programs is selecting tools before aligning leaders, assessing readiness, or defining what success looks like. Phase 1 corrects this by ensuring that every subsequent decision rests on a foundation of shared understanding and explicit agreement.

### Weeks 1–2: Leadership Alignment

Convene a working session with the six to ten leaders who will shape your AI program. This group must span business, technology, risk, compliance, and human resources. Not a steering committee that meets quarterly, but a working group that will make decisions over the next ninety days.

The session has three objectives. First, establish a shared definition of what people-centric AI means for your institution. The definition from Chapter 2 provides a starting point, but the language must be your own, something your leaders can articulate without consulting a slide deck. Second, agree on boundaries: what AI will and won't be used for in the near term, what level of human oversight is non-negotiable, and what risk tolerance the organization is prepared to accept during the pilot phase. Third, name an accountable executive, not a committee, not a working group, but a single person who owns the ninety-day program and has the authority to make decisions when the group can't reach consensus. The right person is typically a Chief Data Officer, Chief Digital Officer, Head of AI or Innovation, or a senior business line leader with cross-functional credibility. Avoid defaulting to the CTO alone; this signals a technology project rather than an organizational transformation, and it limits the buy-in you need from business, risk, and compliance.

This last point is critical. McKinsey's State of AI research has repeatedly linked transformation success to clear executive ownership. Programs owned by committees move at the speed of consensus. Programs owned by a named leader move at the speed of decision.

In parallel, begin the technical groundwork. Conduct a formal kickoff with executive sponsors, technical leads, compliance officers, and business stakeholders. Map the complete stakeholder landscape and establish communication cadences: weekly executive updates, bi-weekly technical deep-dives, and daily channels for the implementation team.

## Weeks 2–3: Readiness Assessment

With leadership aligned, conduct a rapid readiness assessment across four dimensions: data infrastructure, talent and skills, governance maturity, and cultural readiness. The frameworks from earlier chapters (particularly the AI Readiness Landscape from Chapter 1 and the AI Fluency Framework from Chapter 7) provide the assessment structure.

Be honest. The purpose of this assessment isn't to produce a reassuring score for the board; it's to identify the specific gaps that will derail your pilot if left unaddressed. Most financial institutions find that their technology is more ready than their governance, and their governance is more ready than their culture. This pattern is predictable, and addressing it starts with naming it.

In parallel, conduct the technical environment assessment. Audit existing data pipelines, systems, and integration touchpoints. Map regulatory requirements to your solution architecture: SEC, FINRA, OCC, GDPR, and any institution-specific mandates. Engage the compliance and security teams early. This compliance-first design principle ensures that every technical decision is validated against the regulatory framework before implementation begins.

The readiness assessment should take no more than two weeks. If it takes longer, you're over-engineering it. Remember: you're building a foundation for a pilot, not a comprehensive enterprise assessment. The pilot itself will reveal more about your readiness than any assessment ever could.

## Weeks 3–4: Use-Case Selection and Stakeholder Mapping

Select one to three pilot use cases. The selection criteria should balance four factors: business impact, feasibility, people impact, and learning value. A use case with high business impact but low learning value might generate quick savings but teach the organization nothing about how to scale. A use case with high learning value but low business impact might build capability but struggle to maintain executive sponsorship.

The ideal first pilot has moderate complexity, clear success metrics, a willing business owner, and direct impact on either employees or customers, so you can measure people-centric outcomes from the start. Payment exception processing, document classification, customer inquiry routing, and regulatory report generation are common first pilots in financial services because they meet these criteria. To make this concrete, consider what a payment exception pilot looks like in practice. A mid-sized bank processes thousands of payment exceptions daily, transactions that fail automated clearing due to mismatched account details, formatting errors, or missing reference codes. Today, operations analysts manually review each exception, research the cause, and either repair and resubmit or return the payment. An AI pilot would train a model on twelve months of resolved exceptions, deploy it in observe-only mode for two weeks to validate accuracy, then move to a co-pilot mode where the AI recommends a resolution and a human analyst confirms or overrides. By day 90, the team has hard data: the AI's accuracy rate on each exception type, the average time saved per case, the override rate that reveals where human judgment still adds the most value, and the staff feedback on how the tool changes their daily work. That is the kind of bounded, measurable evidence a ninety-day pilot should produce.

For AI systems that will operate with any degree of autonomy, this is also when you define operational boundaries. Using the trust boundary framework from Chapter 8, specify which actions the system can take independently, which require human approval, and which remain human-only. Map each AI action to your institution's existing approval workflows. Document what happens when something goes wrong: how activities are halted, what the impact scope is, and who has authority to intervene. These decisions feel premature at this stage. They're not. Defining boundaries now prevents arguments later, when the pressure to expand the system's scope will be intense.

Simultaneously, map your stakeholders. Identify champions who will advocate for the initiative, skeptics whose concerns must be addressed, and influencers whose support, or opposition, will shape adoption. The change management research is clear: early engagement with skeptics is more valuable than additional time with champions. Skeptics who are heard and involved become your most credible advocates.

If the AI system will observe and learn from live data, deploy it in **observe-only mode** before Phase 1 ends. The system processes real inputs and generates recommendations, but takes zero autonomous actions. This creates the evidence base for every subsequent decision about the system's capabilities: how often does it agree with what humans actually did? Where does it struggle? What surprises the team? This baseline data is invaluable, and it can't be manufactured after the fact.

## *Deliverable: The AI Charter*

Phase 1 concludes with a one-page AI Charter, a document that captures the working group's agreements in plain language. The charter should include: your institution's definition of people-centric AI, the agreed principles and boundaries, the selected pilot use cases with named business owners, the accountable program leader, the governance commitments for the pilot, and, if the system will operate with any autonomy, the operational boundaries and human oversight model. One page. If it requires more, you haven't made enough decisions.

| Element | Instructions |
|---|---|
| **Our Definition of People-Centric AI** | In your own words: what does people-centric AI mean for this institution? |
| **Principles and Boundaries** | 3–5 principles that guide all AI decisions. Include explicit boundaries: what AI will NOT be used for. |
| **Pilot Use Cases** | 1–3 use cases with named business owners and expected outcomes |
| **Accountable Leader** | Name, title, authority, and reporting line |
| **Human Oversight Model** | Where AI acts independently, where it needs approval, where humans decide with AI input |
| **Governance Commitments** | Oversight cadence, review structure, escalation path, documentation requirements |
| **Success Metrics** | Key metrics across business, people, risk, and learning categories |
| **90-Day Timeline** | Phase milestones with dates and owners |

***Figure 10.1: One-Page AI Charter Template***

*Complete collaboratively with business, technology, risk, compliance, and HR leadership. The charter should fit on one page. If it does not, you have not made enough decisions.*

• • •

## Days 31–60: Design with People at the Center

Phase 2 translates alignment into architecture. The critical discipline here is designing governance, oversight, and change management *into* the pilot, not planning to add them later. Every failed AI initiative I've studied made the same mistake: treating people-centric elements as a later optimization rather than a design requirement.

### Weeks 5–6: Pilot Design

Define the pilot scope with tight, deliberate focus. Scope creep is the most common cause of pilot failure, not because the technology can't handle complexity, but because expanding scope dilutes attention, multiplies stakeholders, and makes success harder to measure. The goal isn't to over-engineer the pilot but to draw a clear boundary around what you are testing and what you are not. A well-scoped pilot answers a specific question: Can this AI system, with these governance controls and this human oversight model, deliver these outcomes for these users?

Design the human oversight model explicitly. Refine the trust boundaries you specified in Weeks 3–4: confirm where the AI operates autonomously, where it requires human review before action, and where humans make the final decision with AI providing recommendations. Document these boundaries and ensure every team member, technical and non-technical, understands them.

Define success metrics across four categories. **Business metrics** measure the operational or financial impact: throughput, accuracy, cost, time savings. **People metrics** measure the human experience: employee adoption, perceived usefulness, customer satisfaction, trust scores. **Risk metrics** measure governance effectiveness: override rates, escalation frequency, bias indicators, compliance adherence. **Learning metrics** measure organizational growth: issues identified and resolved, process improvements generated, knowledge captured. Chapter 11 will expand these into a full measurement framework, but establishing targets now, before the pilot launches, is essential.

In parallel, build the core technical infrastructure. Construct and test data pipelines from source systems. Implement data quality checks: schema validation, completeness scoring, anomaly detection. Establish monitoring dashboards for pipeline health and system performance. Every integration point should be tested end-to-end before the pilot begins.

If the system showed promise in observe-only mode during Phase 1, this is the phase to expand its role carefully. Allow it to propose specific actions, with a confidence score and explanation, but require explicit human approval before any action is taken. Begin measuring how often humans agree with the system's recommendations and how often they override them. If the override rate exceeds 25 percent, pause and diagnose before expanding the system's responsibilities. Every human override should feed back into the system's context to improve future recommendations. This feedback loop isn't just a technical step; it's a trust-building exercise. When people see that their judgment improves the system, they engage more deeply and resist less.

## Weeks 6–7: Change Management and Communication

Develop a communication plan for three audiences: the pilot team, the broader organization, and, where relevant, customers and regulators.

For the pilot team, honesty is paramount. Explain what the AI will do, what it won't do, how their roles will be affected, and how their feedback will be used. The research on organizational change is unambiguous: people don't resist change itself; they resist change that's done *to* them rather than *with* them. Involving the pilot team in design decisions, even small ones, dramatically increases adoption and reduces resistance.

For the broader organization, communicate intent without overpromising. Announce that the institution is piloting AI in a specific area, that the pilot includes governance and oversight safeguards, and that results will be shared transparently. Avoid the twin traps of hype and stealth. Hype creates expectations that pilots can't meet. Stealth creates suspicion when people inevitably find out.

For customers, the principle is transparency without alarm. If the AI will interact with customers directly, explain what it does, how human oversight works, and how customers can reach a human if they prefer. The wealth management vignette from Chapter 9 illustrates what happens when firms skip this step.

## Weeks 7–8: Governance Setup and Fluency Kick-start

Establish the governance structure for the pilot: who reviews the AI's outputs, how often, using what criteria, and with what escalation path. This structure should be proportionate to the pilot's risk level (a document classification pilot needs lighter governance than a credit scoring pilot), but it should be explicit, documented, and operational before the pilot launches.

Simultaneously, launch a targeted fluency program for the pilot team and their managers. This isn't a generic AI awareness session; it's a focused workshop on the specific AI system being deployed, how it makes decisions, what its known limitations are, and how to exercise effective oversight. The AI Fluency Framework from Chapter 7 provides the structure: the pilot team needs to operate at the Application tier, understanding enough about the system to use it effectively, challenge its outputs, and identify when something is wrong.

Conduct a formal phase gate review before proceeding. The review should assess technical readiness, governance readiness, and people readiness. Include the compliance team's assessment. If the system has been operating in a co-pilot mode, include accuracy metrics, override rate trends, and any guardrail triggers. Do not proceed to Phase 3 unless all stakeholders are satisfied that the pilot is ready for real-world operation.

### *Deliverable: The Pilot Blueprint*

Phase 2 concludes with a pilot blueprint that integrates scope, metrics, governance, change management, and fluency in a single document. The blueprint is your insurance policy: it ensures that every element needed for a people-centric pilot is in place before the first transaction is processed.

• • •

## Days 61–90: Execute, Learn, and Decide

Phase 3 is where theory meets reality. The pilot launches, real people use the system, real outcomes are measured, and real decisions must be made.

### Weeks 9–10: Controlled Launch

Launch the pilot with a defined cohort: a specific team, a specific workflow, and a specific customer segment. Resist the temptation to expand the cohort during the pilot period. Expansion introduces variables that make it harder to attribute outcomes to the AI system versus other changes happening simultaneously.

Activate the monitoring cadence established in the blueprint. Hold daily check-ins with the pilot team during the first week, transitioning to bi-weekly check-ins thereafter. These aren't status meetings; they're **listening sessions**. The most valuable data in any pilot comes not from dashboards but from the people using the system. What surprises them? What frustrates them? What do they trust? What do they override? These qualitative signals are as important as the quantitative metrics.

For systems operating with any degree of autonomy, implement continuous monitoring: automated alerts when the system's behavior deviates from established baselines, when confidence scores drop, or when patterns change in unexpected ways. If the system has earned enough trust through Phases 1 and 2, carefully expand its responsibilities to low-risk, reversible actions (internal data enrichment, analytics generation, checklist pre-population) while maintaining human approval for anything client-facing, financially material, or irreversible. Document the evidence behind every expansion of the system's role: accuracy rate, override frequency, compliance assessment, and stakeholder sign-off.

Expect problems. Every pilot encounters them. The value of a ninety-day bounded pilot is that problems surface early, when they're cheap to fix, rather than after scale-up, when they're expensive to remediate. Organizations that pilot with structured feedback loops scale far more reliably than those that deploy and hope. In my experience, it's the single clearest dividing line between pilots that graduate and pilots that stall.

## Weeks 11–12: Measure, Learn, and Decide

In the final two weeks, compile the evidence. This is where the four-category metrics framework from the pilot design phase proves its value. Business metrics tell you whether the AI works. People metrics tell you whether the organization is ready. Risk metrics tell you whether governance is effective. Learning metrics tell you what to do differently next time.

Present the results to the working group with a clear recommendation: **scale, pivot, or stop.** Scale means the pilot achieved its objectives across all four metric categories and the organization should proceed to broader deployment with the governance model validated; confirm the scaling sequence and the owner for each wave. Pivot means the pilot revealed that the use case, the technology, or the operating model needs adjustment before scaling; specify what needs to change and propose a revised timeline. Stop means the evidence doesn't support continued investment in this use case; document the reasons and redirect resources.

Stopping isn't failure. It's discipline. The most expensive AI programs in financial services aren't the ones that were stopped; they're the ones that should have been stopped but continued because nobody had the evidence or the courage to make the call. A well-documented decision to stop after a rigorous pilot is worth more to the organization than an undocumented decision to continue based on hope.

Complete the operational handoff. Deliver a comprehensive operations guide covering daily monitoring procedures, weekly review cadences, escalation protocols, and, critically, the process for expanding or constraining the system's autonomy as evidence accumulates. Conduct knowledge transfer sessions covering system architecture, operational procedures, and troubleshooting. Establish ongoing support agreements and define the transition from implementation support to business-as-usual operations.

### *Deliverable: The Pilot Report*

The pilot report is the capstone of the ninety-day program. It includes: a summary of what was tested and how, quantitative results across all four metric categories, qualitative findings from the pilot team and stakeholders, governance observations, lessons learned, and a clear recommendation with supporting evidence. Share the report broadly. Transparency about what worked and what didn't builds organizational trust in the AI program, and organizational trust is the currency you'll need to scale.

• • •

## Framework: The People-Centric Pilot Scorecard

Use the following balanced scorecard to evaluate pilot results across all four dimensions that matter for people-centric AI. Set targets during Phase 2. Record actuals during Phase 3. A successful pilot shows acceptable performance across **all four categories**, not just business metrics. Use the scorecard to support the scale, pivot, or stop decision.

| Category | Metric | Target | Actual |
|---|---|---|---|
| **Business** | Processing accuracy, throughput, cost per transaction, time savings | [Set pre-pilot] | [Record post-pilot] |
| **People** | Adoption rate, user satisfaction, customer trust score, perceived usefulness | [Set pre-pilot] | [Record post-pilot] |
| **Risk** | Override rate, escalation frequency, bias indicators, compliance adherence | [Set pre-pilot] | [Record post-pilot] |
| **Learning** | Issues resolved, process improvements, knowledge captured, feedback quality | [Set pre-pilot] | [Record post-pilot] |

***Figure 10.2: People-Centric Pilot Scorecard***

• • •

## Framework: The 90-Day Planner

The following table provides a week-by-week summary. Customize the specific activities based on your institution's context, but maintain the governance rhythm, the milestone structure, and the phase gates.

| Week | Phase | Key Activities | Milestone |
|---|---|---|---|
| **Week 1** | Assess | Leadership alignment, stakeholder mapping, kickoff | Signed project charter |
| **Week 2** | Assess | Readiness assessment, technical environment audit | Readiness report, gap analysis |
| **Week 3** | Assess | Use-case selection, compliance mapping, boundary definition | Compliance mapping, oversight model |
| **Week 4** | Assess | Architecture design, observe-only deployment, Phase Gate #1 | AI Charter, baseline metrics |
| **Week 5** | Design | Pilot scope, human oversight design, data pipeline construction | Pipeline operational |

| Week | Phase | Key Activities | Milestone |
|---|---|---|---|
| **Week 6** | Design | Four-category metrics, co-pilot testing, data quality checks | Metric targets, override analysis |
| **Week 7** | Design | Governance setup, fluency program, communication plan | Governance operational, training begun |
| **Week 8** | Design | UAT, stakeholder communications, Phase Gate #2 | Pilot Blueprint |
| **Week 9** | Execute | Controlled launch, daily listening sessions, monitoring | Initial performance data |
| **Week 10** | Execute | Expanded operation, continuous monitoring, supervisor training | Mid-pilot review |
| **Week 11** | Execute | Full deployment, operations manual, knowledge transfer | Operational handoff |
| **Week 12** | Execute | Four-category measurement, final review, recommendation | Pilot Report, formal acceptance |

***Figure 10.3: Twelve-Week Execution Summary***

• • •

## Talking to Employees, Customers, and Regulators

Communication isn't a phase of the roadmap; it runs through all three phases. But the stakes are high enough to warrant specific guidance for each audience.

**Employees** need three things: honesty about what is changing, evidence that the organization is investing in them and not just in technology, and a voice in the process. The most effective employee communication I've seen follows a simple formula: here's what we're doing, here's why, here's how it affects you, and here's how you can shape it. Omit any element and trust erodes. The operations transformation at Meridian from Chapter 9 failed precisely because the first deployment communicated the "what" and "why" but not the "how it affects you" or "how you can shape it."

**Customers** need transparency proportionate to impact, though in practice most first pilots are internal and non-customer-facing, which simplifies this step considerably. Where the pilot does touch customers, the principle applies in full. If AI is processing documents behind the scenes, a general privacy notice may suffice. If AI is influencing lending decisions, investment recommendations, or fraud determinations, customers deserve to know, and to have access to a human alternative. The emerging regulatory consensus across jurisdictions is clear: customers must be informed when AI materially affects decisions about them. Getting ahead of this requirement is both an ethical choice and a competitive advantage.

**Regulators** respond well to proactive engagement. Institutions that approach their supervisors with a clear governance framework, documented oversight mechanisms, and transparent reporting earn goodwill and reduce the risk of adversarial examinations. The Pacific Coast vignette from Chapter 9 demonstrates the cost of reactive engagement and the benefit of building governance that regulators can see and evaluate. Several supervisory authorities, including the Monetary Authority of Singapore and the European Banking Authority, have encouraged institutions to engage early and transparently about their AI governance practices.

• • •

## What If You're Not a Big Bank?

I can hear the objection from here, because I've heard it in every workshop I've run for community institutions: "This is all very well for a bank with an AI council, embedded risk partners, and a data science team. We have two hundred employees. Our 'technology function' is six people and a core-systems vendor. Which parts of this chapter are actually for us?"

All of it. Just smaller.

Start with what doesn't change. The four questions behind every framework in this book cost nothing to ask: who is accountable, what the AI can decide alone, how you would know if it went wrong, and what the affected customer can do about it. They scale down perfectly. A credit union deploying a vendor's loan-decisioning tool needs answers to those questions just as surely as a global bank does; the answers simply involve fewer people and shorter documents. Regulators understand proportionality (a community institution will never be held to a money-center bank's governance apparatus), but no examiner anywhere waives accountability because the institution is small.

Now the adaptations. You don't need an AI council; you need AI as a standing agenda item on the risk or management committee you already run, with a one-page charter saying who decides what. You don't need a portfolio of pilots; you need one use case, and for most community institutions it will be a vendor's AI, which means your ninety days lean heavily on Chapter 8's vendor discipline and the due-diligence checklist in Appendix F. The five questions from that chapter are your data science team: a vendor who can't show you validation evidence, change-notification terms, and a kill switch on your side of the boundary is asking a two-hundred-person institution to carry risk that institutions a hundred times larger refuse to carry. Say no. There will be another vendor.

Your fluency program is five people, not five hundred: the CEO, the head of lending or operations who owns the use case, the compliance officer, the person who will actually watch the monitoring reports, and whoever talks to your examiners. Train them first; the rest can follow. Your monitoring infrastructure is whatever the vendor contractually delivers plus one person who actually reads it weekly, which, candidly, puts you ahead of some institutions fifty times your size. And your kill switch can be a documented manual procedure, as long as it's tested and the person who owns it knows they own it.

Here's what the small institution should never forget: you hold the advantage that this entire book is trying to teach large institutions to recreate. Your customers know your people. Your decision chain is three steps, not thirteen. When a member calls about a declined transaction, someone who knows them can look at it today. Big banks spend millions building contestability mechanisms to simulate what you already are. People-centric AI isn't a big-bank luxury you must stretch to afford; it's your home turf, with better tools. The institutions that should worry are the ones that buy an algorithm and abandon that ground.

The ninety-day roadmap, scaled to a community institution, looks like this: Phase 1 is one committee agenda, one inventory (you'll find more embedded AI in your vendor stack than you expect), and one chosen use case. Phase 2 is the vendor diligence and the five-person fluency push. Phase 3 is a supervised go-live with the vendor's monitoring flowing to one named owner. The calendar is the same. The principle is the same. Only the head-count is different, and the trust you're protecting was always your most valuable asset anyway.

• • •

## The Ninety-Day Mindset

The roadmap in this chapter is a tool. But the deeper message is about mindset.

Financial institutions are accustomed to long planning horizons, extensive committee reviews, and gradual consensus-building. These habits have served the industry well in managing risk over decades. But they're poorly suited to AI, where the technology evolves faster than any planning cycle, where organizational learning requires doing rather than analyzing, and where competitive advantage accrues to institutions that build evidence while others build slide decks.

Ninety days isn't a shortcut. It's a discipline. It forces alignment in weeks rather than months. It demands design decisions rather than deferral. It produces evidence that no amount of strategy consulting can replicate. And it builds the organizational muscle (governance, oversight, change management, measurement) that scales far beyond any single pilot.

Start with alignment. Design with people at the center. Execute with discipline. Learn with honesty. Decide with evidence. That's the roadmap, and it's how you keep the human in the lead from the very first day. The next ninety days are yours.

A successful pilot is a start. Chapter 11 addresses the question every executive will ask next: how do we measure the return on all of this?

CHAPTER 11

# Measuring People-Centric ROI

*A CFO's challenge: "The AI team tells me the model saves us two million dollars a year in processing costs. But we've had three customer complaints about unfair decisions, two staff grievances about job changes, and a regulatory inquiry. Is this a good investment?"*

It's the kind of question that exposes the gap between what most organizations measure and what actually matters. The processing savings are real. But so are the complaints, the grievances, and the supervisory letter sitting on the Chief Risk Officer's desk. If the only metric on the dashboard is cost reduction, the executive committee sees a success story. If you expand the lens to include trust, engagement, and regulatory standing, the picture becomes far more nuanced, and far more honest.

This chapter makes the case that traditional return-on-investment frameworks, while necessary, are dangerously incomplete when applied to AI in financial services. We'll build a measurement approach that captures the full spectrum of value, and the full spectrum of risk, that AI creates across your institution.

• • •

## The Limits of Traditional ROI

Every technology investment in banking eventually faces the same question: was it worth it? For decades, the answer has been framed in familiar terms: cost reduction, speed improvement, error reduction, revenue uplift. These metrics are comfortable because they're quantifiable and because they map neatly onto the language finance leaders already speak.

AI initiatives, however, resist this neat framing. Consider a mortgage decisioning model that reduces processing time by 60 percent and eliminates three full-time roles. On a traditional business case, this is an unambiguous win. But what if the model also produces systematically different approval rates across demographic groups? What if the staff whose roles were eliminated had deep institutional knowledge that's now lost? What if customers whose applications were declined have no meaningful way to understand or challenge the decision?

The cost savings are real. The risks are also real. And the risks compound in ways that a spreadsheet built around headcount reduction will never capture.

### *The Efficiency Trap*

The core danger of traditional ROI in AI programs is what I call *the efficiency trap*: the tendency to optimize for the metrics you can easily measure while neglecting the outcomes that matter most. When cost-per-transaction is the primary success metric, every incentive drives toward automation and throughput. Customer experience, employee morale, fairness, and regulatory readiness become secondary considerations, not because anyone decided they don't matter, but because they were never on the scorecard.

Industry research keeps confirming this pattern: among firms that report measurable ROI from AI, nearly all measure operational efficiency, while only a minority track employee-experience metrics, and fewer still systematically measure customer trust. The result is an information gap that makes it impossible for leadership to see whether AI is genuinely creating value or simply redistributing costs from one part of the institution to another.

A narrow ROI view also creates perverse incentives. When the AI team's success is measured solely by efficiency gains, there's no organizational motivation to invest in explainability, fairness testing, or human oversight design. These activities appear as costs on the ledger, never as value creators. Yet they're precisely the capabilities that determine whether an AI initiative survives regulatory scrutiny, maintains customer trust, and achieves sustainable adoption.

• • •

## Employee-Centric Metrics

The first dimension missing from most AI business cases is the impact on the people who work with these systems every day. As we explored in Chapter 3, AI reshapes roles, workflows, and career trajectories. Measuring only whether the AI produced the expected efficiency gains, without measuring how employees experience the change, is like evaluating a surgical procedure by looking only at the incision time while ignoring whether the patient recovered.

### *Adoption and Usage*

The most basic employee metric is also the most revealing: are people actually using the AI tools? Shadow processes (where staff quietly work around the AI system by reverting to manual methods or maintaining parallel spreadsheets) are a reliable indicator that something has gone wrong. High adoption rates suggest the tool is genuinely useful. Low adoption, or declining adoption after an initial spike, signals a gap between what the technology delivers and what the workforce needs.

Track adoption not just as a binary yes-or-no, but as a spectrum: how frequently are people using the tool, which features do they rely on, and which do they avoid? These patterns reveal where the AI adds genuine value and where it creates friction.

### *Perceived Usefulness*

Beyond raw usage, perceived usefulness captures whether employees believe AI makes them more effective at their jobs. This is a subjective measure, typically gathered through surveys and interviews, but its predictive power is substantial. Research in technology adoption consistently shows that perceived usefulness is the strongest predictor of sustained engagement. If frontline advisors view the AI as a helpful colleague rather than an imposed burden, adoption will sustain itself. If they see it as an obstacle or a threat, no amount of executive sponsorship will overcome the resistance.

### *Engagement and Skill Growth*

Two additional employee metrics deserve a place on every AI scorecard. First, *engagement and satisfaction*: has AI improved or worsened the employee experience? This isn't the same as perceived usefulness; an employee might acknowledge that the AI is useful while also feeling anxious about their future role. Track sentiment over time, with particular attention to the transition period when roles are being redesigned.

Second, *skill growth*: are employees developing new capabilities through AI-augmented work? The promise of people-centric AI is that it frees humans for higher-value activities. Measuring whether employees are actually moving into those activities (taking on more complex cases, developing advisory skills, engaging in governance and quality review) validates whether the promise is being kept.

• • •

## Customer-Centric Metrics

If employees are the first people affected by AI, customers are the ultimate beneficiaries or victims. And customers in financial services aren't forgiving. Consumer research in the United Kingdom underscores the point: trust in AI-driven financial decisions remains fragile, and concerns about transparency and fairness persist, themes the FCA itself has highlighted in its work on AI and consumer outcomes. When trust erodes, it doesn't recover quickly. Every AI deployment that touches customers must therefore be measured not just by what it delivers to the institution, but by how customers experience it.

### *Trust Scores*

Trust is the currency of financial services. Measuring it requires asking customers directly: do you believe your institution is using AI responsibly and in your interest? Do you understand how AI-driven recommendations or decisions are made? Do you feel you have meaningful recourse when something goes wrong?

Trust scores should be tracked longitudinally (before, during, and after AI deployment) and segmented by customer type. Wealth management clients, retail banking customers, and small business owners may respond to AI very differently. What matters isn't a single aggregate number but the trajectory and the variation across segments.

### *Experience Quality*

Net Promoter Score, customer satisfaction ratings, and complaint resolution quality are familiar metrics, but they take on new meaning in the context of AI. A virtual assistant that resolves 80 percent of inquiries without human intervention looks efficient on paper. But if the remaining 20 percent of customers who need human help face long wait times or frustrating handoffs, the net customer experience may be worse than before the AI was deployed. Measure the experience holistically: not just the cases AI handled successfully, but the end-to-end journey including escalation and resolution.

### *Vulnerable Customer Outcomes*

Financial regulators globally are placing increasing emphasis on vulnerable customer outcomes. AI systems must be evaluated specifically for how they serve, or fail to serve, customers who are elderly, financially stressed, digitally excluded, or dealing with life events such as bereavement or illness. If an AI chatbot reduces average handling time but makes it harder for a distressed customer to reach a human advisor, the efficiency gain comes at an unacceptable cost.

Track outcomes for vulnerable customer segments separately. This isn't just a regulatory expectation; it's a test of whether your AI deployment is genuinely people-centric or merely cost-centric.

### *Channel Choice*

A subtle but important metric is *channel choice*: are customers using AI-powered channels because they genuinely prefer them, or because the institution has made it difficult to reach a human? The former signals success; the latter signals a forced migration that erodes trust. Monitor the reasons customers escalate from digital to human channels. If the dominant reason is frustration rather than complexity, the AI experience needs redesign.

• • •

## Risk and Regulatory Metrics

The third dimension of people-centric ROI addresses the governance and risk outcomes that traditional business cases often treat as externalities. In financial services, where regulatory scrutiny of AI is intensifying, these metrics are not optional; they are existential.

### *Incidents Avoided and Detected*

AI systems in fraud detection, anti-money-laundering, and operational risk should be measured not just by their detection rates, but by the quality and timeliness of their detections. A fraud model that catches 98 percent of fraudulent transactions sounds impressive until you learn that it also generates a 40 percent false-positive rate that overwhelms the investigation team and delays legitimate customer transactions.

Track true-positive rates alongside false-positive rates. Track the time from detection to resolution. And track the incidents that AI identifies *before they cause customer harm: this is the proactive value* that justifies the investment in governance and monitoring infrastructure.

## Model Issues Detected Early

One of the strongest justifications for investing in AI monitoring and human oversight is the ability to catch problems before they cause harm. Drift detection, identifying when a model's performance degrades because the real-world data has shifted, is a key capability. So is bias monitoring: catching emerging fairness issues before they trigger regulatory action or customer harm.

Track the number and severity of model issues identified through monitoring, the time from detection to remediation, and the estimated impact avoided. This creates a compelling narrative: the governance investment is not overhead; it is a risk mitigation engine.

## Supervisory Feedback and Governance Compliance

Regulatory examination outcomes, audit findings, and adherence to internal AI governance policies provide the final layer of risk measurement. A clean regulatory review isn't just the absence of bad news; it's evidence that the institution's governance framework is working. Track supervisory feedback over time, noting trends in examiner expectations and areas of increasing scrutiny.

Internal governance compliance (whether teams are following documentation requirements, conducting required bias audits, and maintaining audit trails) provides an early warning system. If compliance rates decline, it signals that governance processes are too burdensome, poorly understood, or not valued by the teams that must follow them. Any of these root causes requires attention before it produces a regulatory finding.

• • •

## Building a Balanced AI Scorecard

The preceding sections describe a measurement landscape that is richer and more demanding than traditional technology ROI. The challenge for leaders isn't collecting all of this data; it's synthesizing it into a view that supports sound decision-making. The People-Centric AI Scorecard organizes these metrics into four quadrants, each representing a dimension of value that AI must deliver to be genuinely successful. These quadrants expand the four pilot categories from Chapter 10: Business becomes Business Value, People splits into Employee Experience and Customer Outcomes, Risk becomes Risk and Governance, and Learning runs through all four.

| Quadrant | Metric Category | Example Metrics |
|---|---|---|
| Business Value | Efficiency & Revenue | Cost per transaction reduction<br>Processing time improvement<br>Revenue per advisor uplift<br>Straight-through processing rate |
| Employee Experience | Adoption & Growth | Tool adoption rate (daily active users)<br>Perceived usefulness score (survey)<br>Employee engagement trend<br>New skill acquisition rate |
| Customer Outcomes | Trust & Satisfaction | AI trust score (customer survey)<br>NPS / CSAT for AI-powered journeys<br>Vulnerable customer outcome index<br>Voluntary digital channel adoption |
| Risk & Governance | Controls & Compliance | Model issues detected pre-harm<br>False positive / negative rates<br>Governance compliance rate<br>Regulatory examination outcomes |

***Figure 11.1: People-Centric AI Scorecard***

The scorecard isn't a static document. It requires baselines established *before* AI deployment so that change can be measured accurately. It requires different reporting cadences for different audiences: the board may review the scorecard quarterly, the AI governance council monthly, and delivery teams weekly. And it requires honest interpretation: a scorecard that shows strong business metrics but declining employee engagement or rising customer complaints isn't telling a success story.

### *Setting Baselines*

The single most important step in AI measurement, and the one most frequently skipped, is establishing baselines before deployment. Without a pre-AI baseline for customer trust, employee engagement, and governance compliance, it's impossible to determine whether observed outcomes are improvements, declines, or noise. Invest in baseline measurement during the pilot design phase described in Chapter 10. The cost is modest; the value for subsequent decision-making is immense.

### *Reporting Cadence and Audience*

Not every audience needs every metric. The board needs a quarterly strategic view: are AI investments delivering balanced value across all four quadrants, and are there emerging risks that require attention? The AI governance council needs a monthly operational view: are specific models performing within acceptable parameters, and are governance processes being followed? Delivery teams need a weekly tactical view: are users adopting the tools, are performance metrics trending in the right direction, and are any issues emerging that require immediate attention?

Designing the reporting cadence and audience mapping is itself a governance exercise. It ensures that the right people see the right information at the right time to make sound decisions.

• • •

## Telling the Full ROI Story

Numbers on a scorecard are necessary but not sufficient. Leaders also need a narrative, a coherent story that explains what the AI investment has achieved, what risks it has introduced, and what the evidence suggests about scaling. The ROI Narrative Template provides a structured format for presenting this story to senior stakeholders.

| Section | Content |
|---|---|
| 1. Executive Summary | One-paragraph summary of the AI initiative, its objectives, and the headline outcome across all four scorecard quadrants. |
| 2. Business Value Delivered | Quantified efficiency gains, revenue impact, and processing improvements with comparison to pre-AI baselines. |
| 3. People Outcomes | Employee adoption, engagement, and skill development results. Customer trust, satisfaction, and vulnerable customer outcomes. |
| 4. Risk & Governance | Model performance, incidents detected and avoided, governance compliance rates, and regulatory feedback. |
| 5. Lessons Learned | What worked, what did not, and what the team would do differently. Include both technical and people-related learnings. |
| 6. Recommendation | Evidence-based recommendation: scale, pivot, or stop, with the reasoning transparently documented. |

***Figure 11.2: ROI Narrative Template***

The narrative template serves a dual purpose. First, it forces the AI team to think holistically about its initiative's impact, rather than cherry-picking favorable metrics. Second, it gives senior leaders the context they need to make informed scaling decisions. A business case that shows strong efficiency gains alongside declining customer trust tells a fundamentally different story from one that shows moderate efficiency gains alongside improving trust, and the scaling decision should reflect that difference.

• • •

## From Measurement to Decision

The purpose of measurement isn't to produce reports. It's to produce better decisions. Every metric on the People-Centric AI Scorecard should connect to a decision: if adoption falls below a threshold, what action do we take? If customer trust declines, what triggers a review? If governance compliance drops, what escalation occurs?

Define these decision triggers in advance. A scorecard without decision triggers is a monitoring exercise. A scorecard with decision triggers is a governance tool. The difference matters enormously when the data tells an uncomfortable story and the pressure is on to scale regardless.

## *The Scaling Decision Framework*

Chapter 10's 90-Day Roadmap ends with a decision point: scale, pivot, or stop. This chapter adds an interim verdict, "investigate," for results too ambiguous to call. The People-Centric AI Scorecard provides the evidence base for that decision. As a practical guide, consider the following thresholds:

- **Scale** when all four quadrants show positive trends relative to baselines, with no unresolved issues in risk or governance.
- **Pivot** when business metrics are positive but people or risk metrics show concerning trends that can be addressed through redesign.
- **Stop** when risk or governance metrics indicate systemic issues that cannot be resolved within acceptable time frames, regardless of business metric performance.
- **Investigate** when metrics across quadrants tell conflicting stories (strong efficiency gains but declining employee engagement, for example), and hold the scaling decision until the conflict is resolved.

The critical discipline is that no single quadrant should dominate the decision. An AI initiative that delivers exceptional business value but erodes customer trust or creates governance gaps isn't ready to scale. The balanced scorecard prevents the efficiency trap from driving decisions that the institution will later regret.

• • •

## Practical Implementation Guidance

Building a people-centric measurement system doesn't require a massive data infrastructure project. Start with what you have and expand over time. The following principles will help you establish a credible measurement practice within your existing capabilities.

- **Start with baselines.** Before deploying any AI initiative, measure the current state across all four quadrants. Even rough baselines are better than none.
- **Use existing channels.** Employee surveys, customer satisfaction programs, and risk reporting frameworks already collect much of the data you need. Integrate AI metrics into these channels rather than creating parallel processes.
- **Automate where possible.** Adoption metrics, model performance data, and governance compliance indicators can often be captured through system logging and dashboards, reducing the manual burden.
- **Be honest about what you cannot measure.** Some outcomes, such as the long-term impact on institutional culture or the cumulative effect on customer trust, are difficult to quantify precisely. Acknowledge the limitations and use qualitative data to supplement quantitative metrics.
- **Report transparently.** Resist the temptation to present only favorable metrics. Stakeholders who discover they were given an incomplete picture will lose trust in the measurement system itself, and that trust is difficult to rebuild.

• • •

## Chapter Takeaways

- **Traditional ROI captures only part of the picture.** Trust, engagement, fairness, and governance readiness are real economic outcomes that determine whether AI investments succeed beyond the pilot.
- **Measure what you want to improve.** If you only measure cost, you will only optimize for cost. A balanced scorecard across business, employee, customer, and risk quadrants ensures that people-centric outcomes receive the same leadership attention as efficiency.
- **Baselines are non-negotiable.** Without pre-deployment measurement, it is impossible to determine whether AI is genuinely improving outcomes or simply shifting costs and risks.
- **A scorecard with decision triggers is a governance tool.** Define the thresholds that trigger scaling, pivoting, stopping, or investigating, and commit to acting on the evidence.
- **The full ROI story includes both the numbers and the narrative.** Leaders need context, not just data, to make sound decisions about AI investments.

• • •

*With the right measurement in place, we can face the future with confidence. Chapter 12 looks ahead to agentic AI and the leadership challenge it presents, because the next wave of autonomous AI systems will test every governance framework, every measurement system, and every leadership instinct we've built. Keeping the human in the lead is about to get harder, and more important than ever.*

CHAPTER 12

# Leading Through the Next Wave

## *Agentic AI and Beyond*

*In 2024, your AI answered questions. In 2025, it drafted documents. In 2026, it is booking meetings, filing regulatory reports, and making trade decisions. Where does this end, and who's in charge?*

The question is no longer hypothetical. Across financial services, AI systems are crossing a threshold from tools that assist to agents that act. They don't wait to be asked. They initiate workflows, coordinate with other systems, make decisions within defined boundaries, and execute transactions in milliseconds. This shift, from reactive AI to agentic AI, represents the most consequential technology transition the industry has faced since the arrival of electronic trading.

Everything we've built in this book (the principles, the governance frameworks, the measurement systems, the people-centric design philosophy) has been preparation for this moment. The final chapter asks the question that will define the next era of financial services leadership: how do you stay in the lead when the machines are moving faster than any human can follow?

• • •

## What Agentic AI Means for Finance

Agentic AI refers to systems that can pursue complex goals autonomously, breaking them into subtasks, using tools, and making decisions along the way. Unlike earlier generations of AI that responded to prompts or classified data, agentic systems take initiative. They observe conditions, formulate plans, execute actions, and adjust course based on feedback, all with minimal human intervention.

In financial services, this is already operational reality, not distant speculation. Goldman Sachs, whose agents we met in Chapter 1, has been building autonomous trade-accounting agents that reconcile and process transactions, routing exceptions for human review. Lloyds Banking Group is targeting one hundred million pounds in value from enterprise-wide agentic AI deployment. Morgan Stanley has expanded its GPT-powered advisor tools from information retrieval to proactive meeting preparation and client communication drafting. Payment networks are processing routing decisions in under 200 milliseconds, with AI agents selecting optimal paths across multiple clearing systems.

The scale of what's coming dwarfs what has already arrived. KPMG estimates that agentic AI could unlock roughly three trillion dollars in global productivity gains over the coming decade. In KPMG's pulse surveys, nearly every firm says it plans to deploy AI agents, yet as of early 2026 only about one in ten has done so at scale. The gap between ambition and execution remains vast. What's clear is that every institution's strategy now includes agentic capabilities. The differentiation will come not from whether you deploy agents, but from how you govern them.

### *From Co-Pilot to Autonomous Agent*

The evolution follows a recognizable pattern. The first wave of generative AI in finance functioned as a co-pilot: an advisor could ask a question and receive a synthesized answer. The second wave moved to drafting: the AI produced documents, summaries, and analysis that a human reviewed and refined. The third wave, now underway, is delegation: the human defines a goal, and the AI executes a multi-step workflow to achieve it.

The practical implications are profound. A compliance officer who once reviewed individual transaction alerts now oversees an AI system that triages thousands of alerts, investigates the credible ones, drafts suspicious activity reports, and routes them for human sign-off. A portfolio manager who once ran scenario analyses now delegates that entire process to an agent that monitors market conditions, identifies relevant scenarios, runs the models, and presents findings with recommended actions.

In each case, the human has moved from doing the work to supervising the system that does the work. This shift changes everything about what leadership means, what oversight requires, and what governance must deliver.

• • •

## New Ethical and Governance Questions

Agentic AI doesn't merely intensify existing governance challenges; it introduces fundamentally new ones. The frameworks we built in Chapters 5 and 6 remain essential, but they must evolve to address a landscape where AI systems act with increasing autonomy and interact with each other in ways that no human designed.

### *The Accountability Gap*

When an AI agent makes a decision that harms a customer, who's accountable? The business leader who set the strategic intent? The technology team that built and deployed the agent? The risk function that approved the control framework? The vendor that provided the underlying model? In a world of human decisions, accountability lines are imperfect but traceable. In a world of autonomous agents executing multi-step workflows, the chain of causation becomes far harder to follow.

The EU AI Act provides a starting framework by holding deployers of high-risk AI systems accountable for outcomes, regardless of who built the underlying technology. But legal accountability and practical accountability are different things. Institutions need internal frameworks that trace every consequential AI decision back to a named human who understood the boundaries within which the agent was operating and accepted responsibility for those boundaries being appropriate.

### *The Delegation Paradox*

Here lies the central paradox of agentic AI: the more capable the system becomes, the harder it is for humans to oversee it effectively. When an AI assistant answers a factual question, a human can evaluate the answer against their own knowledge. When an AI agent executes a complex, multi-step workflow involving data from multiple sources, analytical models, and real-time market conditions, the human supervisor may lack the information, time, or expertise to meaningfully evaluate what the agent did and why.

This isn't a failure of the human; it's a structural consequence of delegation. The very reason we delegate to AI is that it can process information at speeds and scales that humans can't. But that same capability makes meaningful oversight harder. The Chapter 6 principle of designing systems that surface the right things for human attention becomes even more critical in the agentic context. Agents must be designed to explain their reasoning at key decision points, flag uncertainty, and pause for human input when the stakes exceed defined thresholds.

### *AI-to-AI Interactions*

Perhaps the most novel governance challenge is the emergence of AI-to-AI interactions. When one institution's trading agent interacts with another institution's pricing agent, the combined behavior may produce outcomes that neither institution anticipated nor intended. Cascading failures become possible: an automated sell decision triggers another agent's risk threshold, which triggers a third agent's hedging strategy, producing a market movement that none of the individual systems was designed to create.

This isn't theoretical. Ever since the May 2010 "flash crash," algorithmic trading interactions have repeatedly produced sudden market dislocations. Agentic AI extends this risk beyond trading into lending, compliance, and customer service. Governance frameworks must now account not just for what your own AI systems do, but for how they interact with external systems whose behavior you don't control.

### *Regulatory Evolution*

Regulators are racing to keep pace. The EU AI Act's high-risk provisions, coming into full force during 2026 and 2027 (with Brussels already debating proposals to extend some deadlines), will test every institution's governance readiness. The European Central Bank has signaled that supervisory expectations for AI governance will tighten further. In the United States, the Office of the Comptroller of the Currency and the Securities and Exchange Commission are increasing scrutiny of algorithmic decision-making and model risk management.

The regulatory direction is clear: as AI systems become more autonomous, the governance expectations for the institutions that deploy them will become more demanding. Institutions that have built robust, documented, people-centric governance frameworks will find regulatory engagement manageable. Those that have treated governance as an afterthought will face uncomfortable conversations and potentially costly remediation.

• • •

## Leadership Behaviors That Will Matter Most

Technology transitions of this magnitude are ultimately leadership challenges. The institutions that navigate the agentic AI era successfully will be led by people who cultivate specific behaviors, not as abstract virtues, but as daily practices that shape decisions, culture, and outcomes.

### *Humility*

The first and most important behavior is humility: the honest acknowledgment that no one fully understands these systems, including the people who built them. Large language models and agentic AI systems exhibit emergent behaviors that their creators didn't explicitly program. They can produce outputs that are brilliant and outputs that are dangerously wrong, sometimes in the same conversation. A leader who claims complete understanding of their AI systems is either uninformed or dishonest.

Humility in practice means creating space for uncertainty. It means saying "we don't fully understand this yet" in a board meeting without it being career-limiting. It means funding research into your own systems' behavior rather than assuming that deployment-time testing was sufficient. It means treating every AI incident as a learning opportunity rather than a failure to be minimized.

### *Transparency*

Transparency is the operational expression of humility. It means being open with employees about how AI will change their roles, including the uncomfortable truths. It means telling customers when they're interacting with an AI system and giving them meaningful alternatives. It means engaging regulators proactively, sharing your governance frameworks and inviting scrutiny rather than waiting for the examination letter.

Transparency also means internal honesty about AI performance. Dashboards that only show favorable metrics, reports that omit customer complaints, and presentations that downplay governance gaps are all forms of opacity that eventually produce larger problems. The measurement framework from Chapter 11 is a transparency tool: it creates a shared view of reality across all four dimensions of AI performance.

### *Cross-Functional Collaboration*

Agentic AI can't be governed by any single function. The business defines the intent, technology builds the capability, risk evaluates the controls, compliance ensures regulatory alignment, legal assesses liability, and human resources manages the workforce impact. If these functions operate in silos, critical governance gaps emerge. If they collaborate effectively, the institution can move with both speed and safety.

Collaboration in practice means more than monthly steering committee meetings. It means embedding risk and compliance professionals in AI delivery teams from day one. It means giving business leaders enough technical fluency to challenge technology proposals meaningfully. It means creating forums where junior staff can raise concerns about AI behavior without navigating three levels of hierarchy. The operating model changes described in Chapter 8 become even more essential as AI systems grow more autonomous.

### *Learning Culture*

Every AI deployment will produce surprises. Some will be pleasant: unexpected efficiencies, novel insights, creative solutions to old problems. Some will be unpleasant: bias discovered in production, customer harm, regulatory findings. The institutions that thrive will be those that learn from both categories with equal rigor.

A learning culture requires psychological safety: the confidence that reporting an AI problem will be treated as a contribution, not a career risk. It requires structured after-action reviews for significant AI incidents, with findings shared across the organization. And it requires a long-term perspective: the value of learning from a failed pilot exceeds the value of hiding the failure to protect the business case.

### *Ethical Courage*

The final leadership behavior is the hardest: the willingness to slow down, stop, or reverse an AI deployment when the evidence says it's causing harm. In an environment where competitive pressure is intense and the board is asking for faster AI adoption, the decision to pause a profitable AI initiative because it produces unfair outcomes or erodes customer trust requires genuine courage.

Ethical courage isn't recklessness. It isn't opposing AI adoption or insisting on zero risk. It's the discipline to hold AI initiatives to the same ethical standards you would apply to any other business decision, and the backbone to act on those standards when the pressure to proceed is strongest.

• • •

## The Agentic AI Governance Readiness Framework

To help leaders assess their institution's preparedness for the agentic AI era, I offer a readiness framework organized around five dimensions. Each dimension represents a critical capability that must be in place before autonomous AI systems can be deployed at scale with confidence.

| Dimension | Key Question | Readiness Indicators |
|---|---|---|
| Accountability Architecture | Can every consequential AI decision be traced to a named human accountable for the boundaries? | Documented accountability chains<br>Named owners for each AI system<br>RACI updated for autonomous workflows |
| Oversight Design | Are human intervention points designed for agent workflows, not just model outputs? | Risk-tiered intervention thresholds<br>Agent pause-and-escalate protocols<br>Calibrated review workloads |
| Explainability at Scale | Can agents explain multi-step reasoning to reviewers at the right level of detail? | Decision logs for agent workflows<br>Audience-tailored explanations<br>Traceability from output to inputs |

| Dimension | Key Question | Readiness Indicators |
|---|---|---|
| Interaction Governance | Have you assessed the risks of your AI agents interacting with external systems? | External interaction inventory<br>Cascading failure scenarios tested<br>Guardrails for AI-to-AI workflows |
| Adaptive Monitoring | Does your monitoring detect emergent agent behaviors, not just predefined failure modes? | Anomaly detection on agent behavior<br>Real-time dashboards for agent activity<br>Automated alerts for boundary violations |

***Figure 12.1: Agentic AI Governance Readiness Framework***

Score each dimension on a scale from one to five, where one represents no capability and five represents a mature, tested, and continuously improving capability. An institution that scores below three on any dimension has significant work to do before scaling agentic AI deployments. The framework is deliberately demanding, because the consequences of deploying autonomous agents without adequate governance are severe and difficult to reverse.

• • •

## Why People at the Center Is the Only Sustainable Strategy

Throughout this book, I've argued that people-centric AI isn't a constraint on innovation but the foundation for innovation that lasts. In the agentic era, this argument becomes even stronger.

AI capabilities will continue to accelerate. Models will become more powerful, agents more autonomous, and the decisions they make more consequential. In this environment of relentless technological change, the one constant is that humans set the purpose, define the values, and bear the consequences. An AI agent can optimize a portfolio, but it can't decide whether the optimization should prioritize short-term returns or long-term client welfare. An agent can triage compliance alerts, but it can't determine what level of risk the institution is willing to accept. An agent can personalize customer interactions, but it can't decide whether personalization should respect privacy boundaries that the law doesn't yet require.

These are human decisions. They require human judgment, human values, and human accountability. No amount of technological capability changes this fundamental reality. The institutions that recognize it, that build their AI strategies around human purpose and human oversight, will earn the trust of customers, regulators, employees, and the broader society. The ones that don't will discover, painfully, that efficiency without trust is a brittle foundation.

## *The Competitive Advantage of Trust*

In a market where every institution has access to similar AI capabilities (the same foundation models, the same cloud infrastructure, the same vendor solutions), differentiation will come from how you deploy those capabilities, not which capabilities you deploy. The bank that can demonstrate fair, transparent, and accountable AI will attract customers who value those qualities. The wealth management firm whose advisors are genuinely empowered by AI, rather than replaced by it, will attract and retain the talent that makes the advisory relationship valuable. The institution with clean regulatory examinations and proactive governance will spend its leadership attention on growth rather than remediation.

Trust is a competitive advantage that compounds over time. Every fair decision, every transparent explanation, every well-governed deployment adds to a reservoir of institutional credibility. And in financial services (where customers entrust institutions with their life savings, their mortgages, and their retirement), that credibility is the most valuable asset on the balance sheet.

### *A Commitment, Not a Slogan*

*Human in the Lead* isn't a marketing tagline. It's an operating philosophy that requires daily commitment from every level of the organization. It means the board asks hard questions about AI governance and expects honest answers. It means the technology team designs for oversight and explainability even when it slows development. It means the business resists the temptation to automate away human judgment in the name of short-term efficiency. It means the risk function challenges AI deployments with the same rigor it applies to any other material risk.

Most of all, it means that when the pressure is on, when the competitor has deployed faster, when the board wants results by quarter-end, when the AI team insists that the governance controls are unnecessary overhead, someone in the room has the courage to say: *we do this the right way, or we do not do it at all.*

That someone is you. The leader reading this book. The one who decided that safe, people-first AI isn't just desirable but essential. The one who understands that the technology will keep changing, but the responsibility to lead with integrity never will.

• • •

## Chapter Takeaways

- **Agentic AI is already here in finance.** Autonomous agents are executing trades, filing reports, and making decisions at scale. Leaders must govern these systems, not just deploy them.
- **The delegation paradox requires new governance approaches.** As AI grows more capable, human oversight becomes both more important and more difficult. Design systems that surface the right decisions for human attention.
- **AI-to-AI interactions create novel risks.** Governance frameworks must account for how your systems in-

teract with external systems whose behavior you don't control.

- **Five leadership behaviors define the next era.** Humility, transparency, cross-functional collaboration, learning culture, and ethical courage are the capabilities that will separate successful institutions from the rest.
- **Keeping people at the center is the competitive advantage.** In a market where AI capabilities are commoditized, trust and governance are the differentiators that endure.

• • •

*The future of AI in financial services will be written by leaders who refuse to choose between innovation and responsibility. Who understand that the fastest path is not always the safest, but that the safest path (built on human judgment, shared accountability, and unwavering commitment to the people you serve) is always the most enduring. The technology will keep evolving. The discipline of keeping the human in the lead is what turns that evolution into advantage.*

# Afterword

## *A Letter to the Leader Reading This*

If you have read this far, you already know that AI in financial services is not primarily a technology challenge. It is a leadership challenge. A trust challenge. A people challenge.

You did not pick up this book because you needed convincing that AI matters. You picked it up because you are looking for a way to do it right, to move with purpose and speed while keeping the people you serve, the people you lead, and the institution you steward at the center of every decision.

I want to be honest with you about something: there is no perfect AI strategy. There is no framework in this book, or in any book, that eliminates the uncertainty ahead. The models will surprise you. The regulations will evolve. Your customers' expectations will shift. Your competitors will move in directions you did not anticipate. The technology itself will change in ways that none of us can fully predict.

What you can control is how you lead through that uncertainty. You can choose transparency over opacity. You can choose to measure what matters, not just what is easy. You can choose to build governance structures that earn trust rather than merely check boxes. You can choose to invest in your people, their skills, their confidence, their ability to work alongside AI systems with judgment and integrity intact.

These are not abstract ideals. They are daily decisions. They show up in the Monday morning meeting when the AI team presents a compelling pilot and you ask, "What does the governance plan look like?" They show up when a customer complaint about an AI-driven decision reaches your desk and you treat it as a signal, not an inconvenience. They show up when a junior analyst raises a concern about model fairness and you listen, really listen, rather than dismissing it as an edge case.

• • •

I have spent fifteen years leading technology transformation and AI programs at financial institutions across North America. I have seen initiatives succeed and I have seen them fail. The pattern is remarkably consistent. The ones that fail treat AI as a technology project. The ones that succeed treat it as a people project that happens to involve technology.

The successful leaders I have worked alongside share common traits. They are curious without being naïve. They are ambitious without being reckless. They listen to their technologists but do not abdicate judgment to them. They care about efficiency but refuse to pursue it at the cost of trust. They understand that the scoreboard has four quadrants, not one.

You do not need to be an AI expert. You need to be the kind of leader who asks the right questions, builds the right team, creates the right culture, and holds everyone, including yourself, accountable for doing this well.

• • •

The title of this book is not accidental. *Human in the Lead* is not a description of how AI works. It is a declaration of how we choose to work with AI. It is a commitment that says: no matter how capable these systems become, the purpose is ours to define, the values are ours to uphold, and the consequences are ours to own.

That commitment starts with you. Today. Not when the strategy is perfect. Not when the governance framework is complete. Not when the regulatory landscape is settled. Now, with the resources you have, the team you lead, and the institution you serve.

Start with one conversation. One pilot designed around people, not just processes. One dashboard that measures trust alongside efficiency. One meeting where you ask a question that no one in the room expected.

The technology will keep evolving. Your commitment to leading with integrity, humility, and care for the people your institution serves: that is what endures. That is what earns trust. That is what builds something worth building.

Thank you for reading. Now go lead.

— *Smit Jani*

Baltimore, Maryland

2026

# Appendix A — Glossary of Key Terms

**Agentic AI:** AI systems that can pursue complex goals autonomously by breaking tasks into subtasks, using tools, making decisions, and adjusting course based on feedback with minimal human intervention.

**AI Fluency:** Contextual understanding of what AI can and cannot do, how to evaluate AI outputs, and how to ask the right questions, tailored to one's role and responsibilities.

**AI Governance:** The organizational structures, policies, processes, and controls that ensure AI systems are developed, deployed, and monitored responsibly and in alignment with institutional values, regulatory requirements, and stakeholder expectations.

**Algorithmic Bias:** Systematic and unfair discrimination in AI outputs that arises from biased training data, flawed model design, or unrepresentative assumptions, producing differential outcomes across demographic groups.

**Autonomy Ladder:** A four-level framework (L0 Shadow Mode through L3 Full Autonomy) for progressively expanding an AI system's decision-making authority based on accumulated trust evidence.

**Circuit Breaker:** An automated safeguard that halts AI agent processing when anomalous conditions are detected, such as excessive action volume, high dollar values, low confidence scores, or behavioral distribution shifts.

**Compliance Sandwich:** An architecture pattern where every AI agent action passes through three layers: pre-action compliance validation, execution with audit trail, and post-action verification.

**Containment Rate:** The percentage of customer interactions fully resolved by an AI system without human intervention. Often used as an efficiency metric but criticized when it prioritizes cost reduction over customer outcomes.

**Data Drift:** A change in the statistical properties of the data an AI model encounters in production compared to the data it was trained on, potentially degrading model accuracy and fairness over time.

**Digital Operational Resilience Act (DORA):** An EU regulation requiring financial institutions to manage ICT-related risks, including those associated with AI systems and third-party technology providers.

**EU AI Act:** The European Union's comprehensive regulatory framework for artificial intelligence, classifying AI systems by risk level and imposing proportionate requirements for transparency, oversight, and accountability. High-risk provisions take full effect in 2026–2027.

**Explainability:** The ability to describe an AI system's outputs and reasoning in terms that are understandable to the audience that relies on them, whether that audience is a customer, advisor, risk officer, or regulator.

**Fairness Testing:** Systematic evaluation of AI model outputs across demographic groups and protected characteristics to identify and mitigate discriminatory patterns.

**Generative AI:** AI systems that create new content (text, code, images, or other outputs) based on patterns learned from training data, typically using large language models or similar architectures.

**Guardrail:** A constraint built into an AI system's architecture that prevents specific undesirable behaviors, including boundary guardrails, financial guardrails, compliance guardrails, quality guardrails, and behavioral guardrails.

**Hallucination:** An AI system's production of outputs that are confident but factually incorrect, fabricated, or unsupported by the input data, a key risk in generative AI applications.

**Human-in-the-Command:** An oversight model where a human sets the rules, monitors AI performance, and intervenes on exceptions, appropriate for medium-stakes, high-volume decisions.

**Human-in-the-Lead:** An oversight model where a human defines the strategic intent, ethical boundaries, and accountability framework within which AI operates autonomously, with the human remaining ultimately accountable for outcomes.

**Human-in-the-Loop:** An oversight model where a human approves or rejects every AI output before action is taken, appropriate for high-stakes, low-volume decisions.

**Kill Switch:** A mechanism to halt AI agent activities at graduated severity levels: Pause (temporary halt), Stop (halt with rollback), and Disconnect (sever all production connections).

**Large Language Model (LLM):** A deep learning model trained on vast amounts of text data that can generate, summarize, translate, and reason about natural language, forming the foundation of most current generative AI applications.

**Model Risk Management (MRM):** The discipline of identifying, measuring, monitoring, and controlling the risk that arises from using quantitative models, including AI and machine learning models, in business decisions.

**Model Validation:** Independent review and testing of an AI model's assumptions, methodology, data, and outputs before deployment and on an ongoing basis to ensure it performs as intended.

**Net Promoter Score (NPS):** A customer loyalty metric based on the likelihood that a customer would recommend the institution to others, often used to measure the impact of AI on customer experience.

**Operational Resilience:** An institution's ability to prevent, adapt to, respond to, recover from, and learn from operational disruptions, including those caused by AI system failures.

**People-Centric AI:** An approach to AI design, deployment, and governance that places human judgment, accountability, and relationships at the center, prioritizing trust, fairness, and human outcomes alongside efficiency.

**RACI Matrix:** A framework that defines roles as Responsible, Accountable, Consulted, and Informed for each activity or decision in an AI initiative, ensuring clear accountability and eliminating governance gaps.

**Shadow Mode:** The L0 level of the autonomy ladder where an AI system observes and generates recommendations against live data but takes zero autonomous actions, creating baseline evidence for trust-building.

**Straight-Through Processing (STP):** The automated, end-to-end processing of a transaction or workflow without manual intervention, a common efficiency metric in AI-augmented operations.

**Transparency:** The principle that stakeholders can see what an AI system does, what data it uses, how it reaches its conclusions, and who is accountable for its outputs.

**Vulnerable Customer:** A customer who, due to personal circumstances such as age, health, financial stress, or life events, is especially susceptible to harm from inadequate AI design or service delivery.

# Appendix B — Regulatory Reference Guide

The following table summarizes the key AI-related regulations, supervisory expectations, and guidance documents most relevant to financial services institutions as of early 2026. This is a living reference; readers are encouraged to monitor regulatory developments in their jurisdictions on an ongoing basis.

| Jurisdiction | Regulation / Guidance | Key Requirements |
|---|---|---|
| **European Union** | EU AI Act (Regulation 2024/1689) | Risk-based classification of AI systems. Credit scoring and creditworthiness evaluation of natural persons classified as high-risk; AI for detecting financial fraud expressly exempted; AML monitoring governed by existing financial-crime regulation. Requires human oversight, transparency, and conformity assessments. Phased implementation 2025–2027. |
| **European Union** | EBA Factsheet on AI Act (Nov 2025) | Confirms AI Act complements existing banking regulation. Existing supervisory expectations for model risk, governance, and operational resilience apply. |
| **European Union** | DORA (Digital Operational Resilience Act) | ICT risk management including third-party AI providers. Incident reporting, resilience testing, and information-sharing obligations. |
| **European Union** | ECB Supervisory Expectations | AI governance at board level. Dedicated AI policies and committees expected. Second and third lines of defense must have adequate AI oversight capabilities. |
| **United States** | Treasury Report on AI in Financial Services (2024) | Guidance on responsible AI adoption. Emphasis on model risk management, consumer protection, fair lending, and third-party risk management. |

| Jurisdiction | Regulation / Guidance | Key Requirements |
|---|---|---|
| **United States** | OCC / Fed / FDIC: Model Risk Management (SR 11-7) | Applies to all models including AI/ML. Requires model validation, ongoing monitoring, outcome analysis, and board-level oversight. Refreshed by updated 2026 supervisory guidance extending coverage to AI/ML models. |
| **United States** | SEC Scrutiny of Algorithmic Decision-Making | Increasing focus on AI in trading, advisory, and compliance. Expectations for explainability, conflict-of-interest management, and investor protection. |
| **United Kingdom** | FCA / PRA AI & Machine Learning Guidance | Pro-innovation, principles-based. Emphasis on accountability, transparency, fairness, and consumer outcomes. Consumer Duty applies to AI interactions. |
| **United Kingdom** | Bank of England: AI in Financial Services | Supervisory focus on model risk, operational resilience, and systemic risk from AI concentration. Stress-testing expectations for AI-dependent processes. |
| **Singapore** | MAS FEAT Principles & Veritas Toolkit | Fairness, Ethics, Accountability, and Transparency principles. Practical assessment tools for bias detection, explainability, and governance. |
| **Global** | FSB / BIS / IOSCO Guidance | Cross-jurisdictional coordination on AI risk in financial stability, market integrity, and consumer protection. Risk-based governance as emerging global standard. |

*Note: Regulatory landscapes evolve rapidly. This guide reflects the state of regulation as of early 2026. Consult legal counsel and compliance teams for jurisdiction-specific advice.*

# Appendix C — Template Library

The following table consolidates all frameworks and templates referenced throughout the book. Each template is designed to be adapted to your institution's specific context, scale, and regulatory environment.

| Chapter | Framework / Template | Purpose |
|---|---|---|
| **Ch. 1** | AI Readiness Landscape | 2×2 matrix mapping regulatory pressure against organizational readiness, with strategies per quadrant. |
| **Ch. 2** | Human-in-the-Lead Decision Matrix | Flowchart for choosing the right human oversight model based on risk, volume, and reversibility. |
| **Ch. 2** | AI Accountability RACI Template | Starter template mapping roles for AI design, deployment, monitoring, and incident response. |
| **Ch. 3** | Role Redesign Canvas | Template for mapping current tasks → AI-augmentable tasks → new human-value tasks → skills gap → training plan. |
| **Ch. 3** | Change Readiness Checklist | 10-question assessment for whether a team is ready for AI-augmented workflows. |
| **Ch. 4** | Customer AI Journey Map Template | Mapping touchpoints where AI adds value vs. where human interaction is essential. |
| **Ch. 4** | Escalation Design Checklist | 8 principles for designing effective human escalation from AI interactions. |
| **Ch. 5** | AI Governance Maturity Model | 5-level maturity scale from ad hoc to optimized, with milestones at each level. |

| Chapter | Framework / Template | Purpose |
|---|---|---|
| **Ch. 5** | AI System Classification Template | Risk-tiering framework for categorizing AI systems and applying proportionate governance. |
| **Ch. 5** | Audit-Readiness Checklist | 15-question assessment for whether an AI system is ready for regulatory review. |
| **Ch. 6** | Oversight Design Canvas | Template for mapping decision types → risk levels → oversight models → escalation paths. |
| **Ch. 6** | Reviewer Training Curriculum Outline | 4-module program for training effective AI oversight skills. |
| **Ch. 7** | AI Fluency Framework | 4-tier model (Awareness → Understanding → Application → Leadership) with role-specific learning objectives. |
| **Ch. 7** | 90-Day Fluency Launch Plan | Phased rollout for introducing AI fluency across the organization. |
| **Ch. 8** | People-Centric Architecture Principles | 10 architecture principles for human-centered AI systems. |
| **Ch. 8** | AI Operating Model Canvas | Template for designing organizational structure around AI delivery and governance. |
| App. F | Vendor AI Due-Diligence Checklist (F.9) | Evaluate and govern third-party AI before signature and at every renewal or material model update. See Appendix F. |
| **Ch. 10** | AI Charter Template | One-page template for documenting AI principles, use cases, boundaries, and governance commitments. |
| **Ch. 10** | People-Centric Pilot Scorecard | Balanced scorecard covering business, people, risk, and learning metrics for AI pilots. |
| **Ch. 10** | 90-Day Planner | Twelve-week execution summary with activities, milestones, and phase gates. |

| Chapter | Framework / Template | Purpose |
|---|---|---|
| **App. F** | Autonomy Ladder | Four-level framework (L0–L3) for progressive trust-based expansion of AI system autonomy. See Appendix F. |
| **App. F** | Guardrail Framework | Five-category guardrail architecture (boundary, financial, compliance, quality, behavioral). See Appendix F. |
| **App. F** | Kill Switch Hierarchy | Three-level halt protocol (Pause, Stop, Disconnect) for agentic AI systems. See Appendix F. |
| **App. F** | Agentic Decision Quality KPIs | Six metrics for measuring AI agent decision quality, trust, and autonomy coverage. See Appendix F. |
| **App. F** | Day 90 Autonomy Profile | Expected autonomy levels by capability at end of 90-day program. See Appendix F. |
| **App. F** | Go-Live Readiness Checklist | 14-item checklist for production deployment including agentic-specific criteria. See Appendix F. |
| **Ch. 11** | People-Centric AI Scorecard | 4-quadrant balanced scorecard (Business, Employee, Customer, Risk) with suggested metrics. |
| **Ch. 11** | ROI Narrative Template | Structured format for presenting the full AI ROI story to senior stakeholders. |
| **Ch. 12** | Agentic AI Governance Readiness Framework | 5-dimension readiness assessment for deploying autonomous AI agents at scale. |

# Appendix D — Recommended Reading and Resources

## *Books*

- **Cathy O'Neil, Weapons of Math Destruction**. Essential reading on how algorithmic bias causes real-world harm, with direct relevance to financial services.
- **Ajay Agrawal, Joshua Gans, and Avi Goldfarb, Prediction Machines**. An economic framework for understanding AI's impact on business decisions and competitive strategy.
- **Amy Webb, The Big Nine. Broad perspective** on the corporations and governments shaping AI's future, useful context for strategic leaders.
- **Erik Brynjolfsson and Andrew McAfee, The Second Machine Age**. Foundational analysis of how digital technologies reshape work, productivity, and the economy.
- **Meredith Broussard, Artificial Unintelligence**. A critical examination of the limitations and biases of AI systems, valuable for maintaining healthy skepticism.
- **Stuart Russell, Human Compatible**. A leading AI researcher's argument for building AI systems that are aligned with human values and remain under human control.

## *Industry Reports and Research*

- **NVIDIA, State of AI in Financial Services (annual). Comprehensive survey** of AI adoption, ROI, and challenges across the financial sector.

- **McKinsey & Company, The State of AI (annual). Cross-industry benchmarking** of AI deployment, investment, and impact.
- **World Economic Forum, AI Governance Alliance Reports**. Multi-stakeholder perspectives on responsible AI governance with sector-specific guidance.
- **Bank for International Settlements (BIS), AI and Machine Learning in Financial Services**. Regulatory perspective on systemic risk and supervisory expectations.
- **KPMG, AI in Financial Services (periodic). Practical insights** on agentic AI, governance readiness, and implementation challenges.
- **Financial Stability Board (FSB), Artificial Intelligence and Machine Learning in Financial Services**. Global regulatory coordination perspectives and systemic risk analysis.

## *Regulatory Sources*

- **European Commission, AI Act Official Text and Guidance. Primary source** for EU AI Act requirements, timelines, and interpretation.
- **European Banking Authority (EBA), AI Act Factsheets. Banking-specific interpretation** of how the AI Act interacts with existing financial regulation.
- **US Department of the Treasury, Artificial Intelligence in Financial Services (2024). US government perspective** on responsible AI adoption in finance.
- **Monetary Authority of Singapore, FEAT Principles and Veritas Toolkit. Practical tools** for fairness, ethics, accountability, and transparency in AI.
- **UK Financial Conduct Authority, AI and Machine Learning Guidance. Principles-based approach** to AI governance with consumer protection focus.

## *Organizations and Communities*

- **Partnership on AI. Multi-stakeholder organization** advancing responsible AI practices across industries.
- **AI in Financial Services Forum (OECD). International coordination** on AI policy and governance in the financial sector.
- **The Alan Turing Institute. Leading research** on trustworthy AI, fairness, and explainability with practical tools and frameworks.
- **IEEE Global Initiative on Ethics of Autonomous and Intelligent Systems. Technical standards** and ethical frameworks for AI development and deployment.

# Appendix E — Discussion Guide

The following questions are designed for executive book clubs, leadership team offsites, or board-level discussions. Each set corresponds to a chapter and is intended to provoke reflection, debate, and practical action planning. Select the questions most relevant to your institution's current AI journey.

• • •

## *Introduction — Why This Book, Why Now*

- What is our institution's current definition of "success" for AI? Does that definition include people outcomes, or is it primarily about efficiency?
- If we asked ten leaders at our institution, "Are we falling behind on AI?" would we get ten different answers? What does that tell us about alignment?

• • •

## *Chapter 1 — Why People-Centric AI Matters Now*

- Which quadrant of the AI Readiness Landscape does our institution currently occupy? What would it take to move to a more favorable position?
- What is the single greatest barrier to responsible AI adoption at our institution: regulatory uncertainty, talent, data infrastructure, or culture?

• • •

## *Chapter 2 — Principles of People-Centric AI in Financial Services*

- For our most important AI use cases, which oversight model is most appropriate: Human-in-the-Loop, Human-in-the-Command, or Human-in-the-Lead? Are we applying the right model today?
- Can we clearly articulate who is accountable for the outcomes of each AI system we deploy? If not, where are the accountability gaps?

• • •

## *Chapter 3 — AI and the Future of Work*

- How are we communicating with employees about AI's impact on their roles? Is our communication honest, or are we avoiding difficult truths?
- Do our employees have visible upskilling paths for an AI-augmented world, or are we asking them to trust that it will work out?

• • •

## *Chapter 4 — Designing Customer-First AI Journeys*

- How easy is it for our customers to reach a human when an AI interaction is not meeting their needs? Would we be comfortable with the answer?
- Are we measuring customer trust in our AI-powered services, or only measuring cost-per-interaction?

• • •

## Chapter 5 – Governance Foundations

- If a regulator asked us tomorrow, "Who approved this AI model, how do you know it's fair, and what happens when it goes wrong?" could we answer confidently?
- Does our board have sufficient AI literacy to provide meaningful oversight, or is AI treated as "an IT thing"?

• • •

## Chapter 6 – Human Oversight in Practice

- Are our human oversight processes meaningful, or have they become rubber-stamp exercises? How would we know the difference?
- Do our staff feel safe challenging AI recommendations, or does the organizational culture discourage dissent?

• • •

## Chapter 7 – Building AI Fluency

- What AI fluency level do we need at the board, executive, management, and frontline levels? How far are we from those targets?
- Does our AI training include ethics, bias, and governance, or only how to use the tools?

• • •

## Chapter 8 – Architecture and Operating Models

- Are our AI systems designed for transparency and auditability, or will we need to retrofit these capabilities later?
- Do we have risk and compliance professionals embedded in AI delivery teams, or are they only involved at the approval gate?

• • •

## Chapter 9 — Case Vignettes

- Which vignette resonates most with our institution's experience? What lessons can we apply from that story?
- Have we experienced a technology-first deployment that required a people-centric course correction? What did we learn?

• • •

## Chapter 10 — The 90-Day Roadmap

- Could we draft an AI Charter for our institution in the next 30 days? What would be the hardest part: agreeing on principles, selecting use cases, or naming accountable leaders?
- Is our next AI pilot designed to test governance and change management, or only to test the technology?
- If we were deploying an AI system with any degree of autonomy, do we have the guardrail architecture, kill switch hierarchy, and monitoring infrastructure to do it responsibly? Where are the gaps?

• • •

## Chapter 11 — Measuring People-Centric ROI

- What would our AI scorecard look like if we measured employee experience and customer trust alongside cost reduction? Would the story change?
- Do we have pre-deployment baselines for the metrics that matter most? If not, what is our plan to establish them?

• • •

## *Chapter 12 – Leading Through the Next Wave*

- How prepared are we for agentic AI? If we assessed ourselves against the five dimensions of the Governance Readiness Framework, where would our gaps be?
- Which of the five leadership behaviors (humility, transparency, collaboration, learning culture, ethical courage) is strongest at our institution? Which needs the most development?
- When the pressure is on to deploy faster, who in our organization has the standing and the courage to say, "We do this the right way, or we do not do it at all"?

# Appendix F — 90-Day Implementation Toolkit

This appendix provides the detailed operational frameworks, checklists, and reference tables that support the 90-Day Roadmap in Chapter 10. While the chapter focuses on strategy, governance, and people-centric execution, these tools provide the implementation depth that technical leads, program managers, and governance teams will need to put the roadmap into practice.

## *F.1: The Autonomy Ladder*

Every AI capability that involves any degree of autonomous action should move through these four levels during the ninety-day program. The autonomy ladder is the operational expression of the trust boundary framework from Chapter 8. Movement up the ladder requires documented evidence; movement down is always available when conditions change.

| Level | Agent Behavior | Human Role | Financial Services Example |
|---|---|---|---|
| **L0: Shadow Mode** | Observes and recommends but takes no action | Executes all actions; reviews AI suggestions | AI flags potential compliance issues but analyst reviews all |
| **L1: Co-Pilot** | Drafts actions; human approves before execution | Reviews and approves each AI-proposed action | AI pre-fills trade allocation; trader confirms and submits |
| **L2: Supervised Autonomy** | Executes within guardrails; human reviews outcomes | Monitors dashboards; intervenes on exceptions | AI auto-generates client reports; portfolio manager (PM) reviews before distribution |
| **L3: Full Autonomy** | Executes autonomously within defined boundaries | Sets boundaries and reviews aggregate performance | AI auto-rebalances model portfolios within risk limits |

*Critical rule: Never target L3 by Day 90 for any workflow involving client-facing decisions, trade execution, or regulatory reporting. Aim for L2 as the ambitious target, with a documented post-90-day path to L3 based on trust evidence.*

## F.2: Deliverables by Phase

### Phase 1 Deliverables (Days 1–30)

| Deliverable | Description |
|---|---|
| **AI Charter** | One-page agreement on definition, principles, boundaries, use cases, and governance |
| **RACI Matrix** | Complete responsibility map for all ninety-day workstreams |
| **Technical Design Document** | Architecture, data flows, integration specifications, and deployment strategy |
| **Compliance Mapping Report** | Regulatory requirements mapped to solution controls |
| **Agent Capability Matrix** | Complete list of AI actions with autonomy level targets and approval mappings |
| **Guardrail Architecture** | Hard limits, soft limits, monitoring rules, and kill switch design |
| **Shadow Mode Report** | AI running on live data in observe-only mode with baseline accuracy metrics |
| **Red Team Assessment** | Adversarial testing results with remediation plan |
| **Risk Register** | Identified risks with owners, mitigations, and escalation paths |
| **Phase Gate #1 Report** | Formal assessment of Phase 1 completion and Phase 2 readiness |

## *Phase 2 Deliverables (Days 31–60)*

| Deliverable | Description |
|---|---|
| **Pilot Blueprint** | Integrated scope, metrics, governance, change management, and autonomy plan |
| **Autonomy Expansion Report** | L0 to L1 to L2 transition evidence with accuracy and override metrics |
| **Agent Accuracy Baseline** | AI vs. human decision concordance rate with confidence intervals |
| **Integration Test Results** | End-to-end test evidence for all data pipelines and API connections |
| **UAT Test Plan** | Business workflow test cases with expected and actual results |
| **Human Override Analysis** | Pattern analysis of when and why humans override the AI, with improvement actions |
| **Red Team Assessment #2** | Expanded adversarial testing including multi-agent scenarios |
| **Phase Gate #2 Report** | Formal assessment and production readiness checklist with agentic criteria |

## *Phase 3 Deliverables (Days 61–90)*

| Deliverable | Description |
|---|---|
| **Pilot Report** | Comprehensive results across business, people, risk, and learning metrics with recommendation |
| **Agent Operations Manual** | Complete operations guide including autonomy management procedures |
| **Success Metrics Dashboard** | KPI dashboard showing baseline vs. post-implementation performance |
| **Autonomy Progression Evidence** | Documented L0 through L2 transitions with supporting data for each workflow |

| Deliverable | Description |
|---|---|
| **L3 Autonomy Roadmap** | Post-ninety-day plan with evidence thresholds for full autonomy |
| **Compliance Audit Package** | Complete audit trail, red team results, and regulatory sign-off |
| **Final Acceptance Report** | Formal sign-off documenting all deliverables completed and success criteria met |

## *F.3: The Guardrail Framework*

The guardrail framework addresses the deepest concern every leader, regulator, and employee will raise about AI systems that act with any degree of autonomy: what happens when the system does something wrong?

| Category | What It Prevents | Example Rule | Enforcement |
|---|---|---|---|
| **Boundary** | Acting outside defined scope | Cannot access systems outside approved tool list | Hard-coded in orchestrator; cannot be overridden |
| **Financial** | Excessive financial exposure | No single action exceeds $X; daily aggregate limit of $Y | Pre-action check with circuit breaker |
| **Compliance** | Regulatory violations | Cannot process restricted-list securities; must check blackout periods | Real-time compliance validation |
| **Quality** | Hallucinated or low-quality outputs | All generated content must include source citations | Output validation with source verification |
| **Behavioral** | Unintended behavioral patterns | Alert on repeated similar recommendations within one hour | Statistical monitoring with anomaly detection |

## *F.4: The Kill Switch Hierarchy*

Every deployment of an AI system with autonomous capabilities must include a clearly defined kill switch hierarchy. Three tiers provide a response proportionate to severity.

**Level 1: Pause.** Temporarily halt AI processing. Queued work is held, not lost. Any team member can trigger it. Auto-resumes after review. This is the everyday safety valve for when something looks off but is not yet confirmed as a problem.

**Level 2: Stop.** Halt all AI activities and roll back any in-flight actions. Requires senior operator approval. Manual restart only after root cause analysis. This is for confirmed issues that require investigation before the system should process another request.

**Level 3: Disconnect.** Sever all AI connections to production systems. Only the CTO or CISO can authorize. Full revalidation required before reconnection. This is the emergency brake, reserved for scenarios where the system's behavior threatens material harm.

## *F.5: Architecture Patterns for Regulated Environments*

Two architectural patterns are essential for financial services AI deployments that involve any degree of autonomous action.

**The Compliance Sandwich.** Every AI action passes through a three-layer validation stack. First, a compliance check validates the proposed action against regulatory rules and internal policies *before* execution. Second, the action executes with complete logging of inputs, reasoning, tool calls, and outputs. Third, a verification step confirms the action was completed correctly and the outcome matches the intent. This pattern ensures that no action escapes compliance review, regardless of autonomy level.

**The Circuit Breaker Pattern.** Automated circuit breakers halt AI processing when anomalous conditions are detected. A *volume* circuit breaker halts if the system executes more than a threshold number of actions per time window. A *value* circuit breaker halts if the cumulative dollar value of AI-initiated actions exceeds a threshold. A *confidence* circuit breaker halts if the system's confidence score drops below a defined threshold on consecutive decisions. An *anomaly* circuit breaker halts if the system's decision distribution deviates significantly from its historical baseline. Together, these prevent runaway loops, excessive financial exposure, and model drift.

## *F.6: Agentic Decision Quality KPIs*

Standard software KPIs (uptime, latency, error rate) are necessary but insufficient for AI systems that make decisions with any degree of autonomy. The following metrics measure decision quality, not just availability.

| KPI | Target | Measurement | Phase |
| --- | --- | --- | --- |
| **AI-Human Concordance** | >85% by Day 60 | % of AI recommendations matching human decisions | Phase 1–2 |
| **Human Override Rate** | <15% by Day 90 | % of AI actions overridden by humans | Phase 2–3 |
| **Guardrail Trigger Rate** | <5% of actions | % of actions triggering guardrail intervention | Phase 2–3 |
| **Autonomy Coverage** | >60% at L2+ | % of defined workflows at L2 or above | Phase 3 |
| **Decision Explanation Score** | >4.0/5.0 | User rating of AI reasoning transparency | Phase 2–3 |
| **Time Saved per Analyst** | >3 hrs/day | Before/after time-motion study | Phase 3 |

### *Trust Metrics*

Beyond operational metrics, track the signals that indicate growing organizational confidence. **Autonomy level progression rate**, how quickly workflows move from L0 to L2, is the clearest indicator. **Scope expansion requests**, the number of new workflows stakeholders ask to add, measure organic demand.

**Declining escalation frequency** over time indicates growing trust. And a weekly **satisfaction score** from users who interact with the system captures the human experience that no dashboard can fully represent.

## F.7: Day 90 Autonomy Profile

The following table illustrates a typical Day 90 autonomy profile for a comprehensive AI deployment in financial services. Customize based on your institution's risk appetite and regulatory context.

| Capability | Day 90 Autonomy Level |
|---|---|
| **Data ingestion and enrichment** | L3: Fully autonomous with monitoring |
| **Research and analysis generation** | L2: Auto-generate with human review before distribution |
| **Portfolio analytics and reporting** | L2: Auto-generate with approval workflow |
| **Compliance monitoring and flagging** | L2: Auto-flag with compliance officer review |
| **Trade idea generation** | L1: Co-pilot with explicit trader approval |
| **Trade execution** | L0/L1: Shadow or co-pilot only; L2+ requires extended trust period |
| **Client-facing communication** | L1: AI drafts, human reviews and sends |

## F.8: Go-Live Readiness Checklist

Complete before any workflow moves beyond L1 autonomy. All items must be checked and signed off by the designated owner before proceeding.

| Readiness Item | Status | Owner |
|---|---|---|
| All integration test cases passed | ☐ | [Assign] |
| UAT sign-off from business stakeholders | ☐ | [Assign] |
| AI accuracy baseline meets threshold | ☐ | [Assign] |
| Guardrail architecture tested and operational | ☐ | [Assign] |
| Kill switch hierarchy tested at all three levels | ☐ | [Assign] |
| Red team assessment remediation complete | ☐ | [Assign] |
| Security penetration test remediation complete | ☐ | [Assign] |
| DR/BCP failover test passed | ☐ | [Assign] |
| Operations manual delivered and reviewed | ☐ | [Assign] |
| User training completed for all roles including supervisors | ☐ | [Assign] |
| Monitoring and alerting configured and tested | ☐ | [Assign] |
| Rollback procedure tested and documented | ☐ | [Assign] |
| Compliance sign-off obtained | ☐ | [Assign] |
| Executive go/no-go decision documented | ☐ | [Assign] |

*Instructions: Complete collaboratively with technology, business, risk, and compliance stakeholders. Do not proceed to expanded autonomy until all items are checked. Revisit this checklist whenever the system's autonomy level is expanded.*

## F.9: Vendor AI Due-Diligence Checklist

Use during vendor selection, before contract signature, and again at every renewal or material model update. A vendor AI system that cannot pass this checklist should not be making decisions about your customers. (See Chapter 8, "When You Didn't Build It: Governing Vendor AI.")

| Area | What to Verify | Status |
|---|---|---|
| Transparency & documentation | Intended use and tested limits documented; training data described sufficiently to assess relevance and bias; fairness testing results provided; known failure modes disclosed; documentation refreshed at each material update | ☐ |
| Validation on your data | Pre-production validation on the institution's own data completed; performance thresholds agreed and contractual; ongoing re-validation rights secured; results reviewed by model risk function | ☐ |
| Change management | Advance notice of material model updates (defined in days); version pinning or staged rollout available; re-validation window before updates affect regulated decisions; change log delivered to institution | ☐ |
| Data protection | Data sent to vendor minimized and documented; retention limits contractual; customer data excluded from vendor model training (or explicitly approved); deletion on exit provable; processing locations known and compliant | ☐ |
| Audit & oversight | Audit rights and regulator access contractual; decision logs and lineage delivered in institution's format; performance and drift metrics visible to institution's monitoring; incident notification within defined hours | ☐ |
| Intervention & exit | Kill switch operable from institution's side without vendor involvement; fallback process defined for vendor outage; exit and deconversion plan documented and tested; data portability confirmed | ☐ |
| Accountability mapping | Internal owner named for the vendor relationship and for the model's outcomes; vendor system entered in AI inventory with risk tier; RACI updated; concentration risk assessed against existing vendor footprint | ☐ |

*Instructions: To be completed jointly by the business owner, model risk, and third-party risk management. Any unchecked item requires a documented exception (approved through governance before deployment) and a remediation date.*

# Notes and Sources

*The case vignettes in Chapter 9 and the institutions described as composites throughout this book are drawn from real experiences across multiple financial institutions, anonymized and combined; the names are fictional. The notes below identify the published sources behind specific claims and statistics, keyed by chapter. Where a figure carries no note, it reflects the author's practice experience or industry surveys current as of early 2026.*

## Introduction

AI-agent ambition versus deployment at scale: KPMG, AI Quarterly Pulse Survey series (2025–2026), kpmg.com.

ATMs and teller employment: James Bessen, Learning by Doing: The Real Connection between Innovation, Wages, and Wealth (Yale University Press, 2015); see also Bessen, "Toil and Technology," IMF Finance & Development, March 2015.

EU AI Act: Regulation (EU) 2024/1689, in force August 1, 2024, with phased application through 2027; implementation timeline at the European Commission's AI Act Service Desk, ai-act-service-desk.ec.europa.eu. The exemption of financial-fraud detection from the high-risk creditworthiness category appears in Annex III, point 5(b).

## Chapter 1

Adoption statistics (80 percent active AI use, up from 31 percent; 26 percent agentic AI, with 51 percent of those beyond pilots): Broadridge, 2026 Digital Transformation & Next-Gen Technology Study, broadridge.com.

EBA mapping of the AI Act to EU banking regulation: European Banking Authority, "AI Act: Implications for the EU Banking and Payments Sector" (factsheet), November 21, 2025, eba.europa.eu.

US Treasury: Artificial Intelligence in Financial Services, US Department of the Treasury, December 19, 2024.

Goldman Sachs agents built with Anthropic: CNBC, "Goldman Sachs Taps Anthropic's Claude to Automate Accounting, Compliance Roles," February 6, 2026; American Banker, "Goldman Equips AI Agents to Do Trade Accounting, Onboarding," February 2026.

Lloyds Banking Group agentic AI program and £100 million value target: "2026: The Year of Agentic AI, and a New Era for Finance," lloydsbankinggroup.com.

AI guardrails and incidents (2 percent / 95 percent): Infosys Knowledge Institute, Responsible Enterprise AI in the Agentic Era, August 2025 (survey of 1,500+ senior executives).

Wealth management AI impact (70 percent): Wipro, AI in Wealth Management: Navigating an Evolving Data-Driven Landscape, October 2024 (survey of 100 US executives).

**Chapter 3**

Accenture on agentic AI in KYC and banking operations: Accenture Banking Blog, "Agentic AI and the Future of Work in Financial Services," bankingblog.accenture.com.

**Chapter 4**

Personalization expectations (71 / 76 percent) and the revenue effect of personalization: McKinsey & Company, "The Value of Getting Personalization Right, or Wrong, Is Multiplying" (Next in Personalization), November 2021.

Chatbot failures in banking: Consumer Financial Protection Bureau, Chatbots in Consumer Finance (Issue Spotlight), June 2023.

UK chatbot investigation: Which?, "Can You Trust AI with Your Money?" November 2025.

Banking-chatbot security testing (24 models, 1–64 percent manipulation rates): Milton Leal (TELUS Digital), "I Tested 24 AI Banking Chatbots; They Were All Exploitable," Corporate Compliance Insights, January 2026.

Data-transparency perceptions (63 percent): Salesforce, State of the Connected Customer, 3rd ed., 2019.

Global chatbot market size: Mordor Intelligence, Chatbot Market report, 2026 estimate.

## Chapter 5

ECB workshops with thirteen banks and governance observations: ECB Supervision Newsletter, "AI's Impact on Banking: Use Cases for Credit Scoring and Fraud Detection," November 20, 2025; supervisory priorities 2026–2028 at bankingsupervision.europa.eu.

EBA mapping exercise and "no significant contradictions" finding: EBA factsheet of November 21, 2025 (cited above) and the accompanying EBA Chair letter on the mapping outcome.

## Chapter 6

Human oversight requirements and automation bias: Regulation (EU) 2024/1689, Article 14.

Automation bias in expert readers: Thomas Dratsch et al., "Automation Bias in Mammography: The Impact of Artificial Intelligence BI-RADS Suggestions on Reader Performance," Radiology 307, no. 4 (2023): e222176.

US model risk guidance: Federal Reserve SR 11-7 / OCC Bulletin 2011-12, Supervisory Guidance on Model Risk Management, April 2011; refreshed by updated supervisory guidance in 2026.

## Chapter 8

Multi-agent system demand (1,445 percent inquiry surge) and the 40 percent enterprise-application prediction: Gartner, "Multi-agent Systems in Enterprise AI," gartner.com, and Gartner press release, August 26, 2025.

Guardrails statistic: Infosys (cited in Chapter 1).

Deployer obligations regardless of who built the model: Regulation (EU) 2024/1689.

## Chapter 10

Executive ownership and AI transformation outcomes: McKinsey & Company, The State of AI survey series, mckinsey.com.

## Chapter 11

Perceived usefulness and technology adoption: Fred D. Davis, "Perceived Usefulness, Perceived Ease of Use, and User Acceptance of Information Technology," MIS Quarterly 13, no. 3 (1989).

**Chapter 12**

Morgan Stanley advisor tools: "Launch of AI @ Morgan Stanley Debrief," Morgan Stanley press release, June 26, 2024; the AI @ Morgan Stanley Assistant launched September 2023.

Agentic AI productivity estimate (~$3 trillion): KPMG, "AI's Next Frontier: Unlocking 'Total Value' through Agentic Technologies"; see also World Economic Forum/KPMG, January 2026.

The May 2010 flash crash: CFTC and SEC, Findings Regarding the Market Events of May 6, 2010 (joint report, September 2010).

EU AI Act high-risk implementation dates: Regulation (EU) 2024/1689 (cited in the Introduction).

# Acknowledgments

This book started as an idea and became real because of the people who pushed it forward.

**Sanjay Kukreja, Chief Technology Officer at eClerx and a leader whose career reflects the very transformation discussed in these pages**, agreed to write the foreword. In doing so, he lent this book something a first-time author can't manufacture alone: credibility earned through years of leading organizations through change. Thank you for putting your name behind mine.

That foreword exists because **Charles Zehil, colleague and mentor at eClerx**, pushed for it. His suggestion to circulate an early pre-read within eClerx is what put the manuscript in front of Sanjay in the first place. A book is often shaped as much by the people who open doors as by the people who write inside them.

**Michael Librot, a colleague, a mentor, and a thoughtful reviewer**, gave feedback on individual chapters that sharpened the arguments where they needed it most.

**Carol Ann Mathison, whose insight, generosity, and encouragement greatly influenced this manuscript,** read every page of it, more than once, and returned feedback sharp enough to catch what I could no longer see in my own writing. Any clarity in these chapters owes something to her.

**Rachna Sahni and TJ Panth, former colleagues whose friendship has outlasted our time working together,** read early drafts when the book was far from finished and far from certain, and kept me writing when it would have been easier to stop.

Finally, to my family, my parents, and my cousin Kiran Pandya: thank you for the continuous motivation and support, offered without pause, long before this book was anything more than an idea. None of this happens without you.

# About the Author

Smit Jani has spent fifteen years leading technology transformation at financial institutions across North America, from global banks to wealth managers, guiding leadership teams through the gap between what AI promises in the boardroom and what it takes to make it work on the floor. The frameworks in this book, from the three models of human oversight to the ninety-day roadmap, were built and tested inside those engagements, not theorized from outside them.

He is a management consultant at eClerx, leading an AI transformation and automation squad at a global investment bank. *Human in the Lead* is his first book.

He lives in Baltimore. Connect with him at linkedin.com/in/smit-jani or smit.jani@outlook.com.